I0002719

Moving from 2D to 3D CAD
for Engineering Design

Challenges and Opportunities

Louis Gary Lamit

James Gee - Technical Assistance

Download the Pro/ENGINEER Tryout Edition
and view free online tutorials at:
http://www.ptc.com/offers/tryout.htm

Moving from 2D to 3D CAD for Engineering Design
Challenges and Opportunities

Louis Gary Lamit
James Gee - Technical Assistance

Copyright © 2007 Louis Gary Lamit
All rights reserved.
ISBN: 1-4196-6426-3
Library of Congress Control Number: 2007902376
To order additional copies, please contact us.
BookSurge, LLC
www.booksurge.com
1-866-308-6235
orders@booksurge.com

Dedication

This book is dedicated to Cooper

Om Mani Padma Hum

Scholarships for Veterans (SFV)

A portion of this text's profits go to my scholarship fund at Foothill-De Anza Community College District (FHDA Foundation). We fund one or more scholarships per year. Your contributions provide extra scholarships, as funds are available. Scholarships for Veterans hope to motivate individuals and businesses to assist in providing scholarships for returning veterans. Money donated goes directly into the designated college's scholarship fund and is collected and awarded by the college foundation. SFV provides funding for a 2-year AS, or AA degree, which covers tuition and fees (or applied to expenses) for 90-quarter or 60-semester units up to $2000 (US). Scholarships are available to any qualified veteran of the Army, Navy, Air Force, Marines, or Coast Guard. The local college foundations administer scholarships. No administration fees are taken by Scholarships for Veterans. All costs associated with Scholarships for Veterans are borne by Lamit and Associates and CAD-Resources.com. We encourage you to establish a scholarship at your local community college or university, or contribute to an existing scholarship for veterans. For more information, see www.scholarshipsforveterans.org.

Preface

In 1987 we wrote: "In the near future, a solid model will form the master representation of a part in contrast to the current practice of using the engineering drawing as the master representation. The main output of the design/drafting office will be a solid model of the part together with all the associated information that is contained on the engineering drawing and the provision of the engineering drawing will be a secondary function. In particular, the drawings, if required, will be generated from the model. The combination of a solid model and the necessary tolerance and associated technical data will be called the product model. Functions downstream of the design office will take the product model as their primary input."

CADD: Computer Aided Design and Drafting, by Louis Gary Lamit and Vernon Paige

     A mere 20 years after I wrote this with my coauthor, industry is still engaged in the debate, 2D or 3D, or both. I had spent time employed as a drafter and designer (1966-73), and teaching "drafting" (1973-1984) all "on the board". In 1984 I took a job as the CAD/CAM instructor at De Azna College. I taught traditional drafting and design and descriptive geometry, but my primary class load was teaching Computervsion. All work was done as 3D wireframe modeling and drawings were derived from the model. I thought I was in heaven, until, we added a new "CAD" system called AutoCAD in 1987. It was very difficult for me to understand why we would go backwards and do things similar to the drafting board - drawing in 2D, albeit with the aid of a computer. Almost twenty years later we still teach two classes in 2D CAD using AutoCAD along with classes in 3D CAD: nine in Pro/ENGINEER, five in SolidWorks, two each in Inventor and Unigraphics and one each for AutoCAD 3D Civil, and AutoDesk Architectural Desktop. Though a vast majority of jobs for my students (most are degreed professionals) are in positions using 3D CAD (50% Pro/E and 30% SolidWorks) there is still a need for the use and understanding of 2D CAD (AutoCAD in our case). Many companies still have legacy projects on 2D, and some still use it as their primary design tool. This book is meant to guide and assist individuals and companies in their quest for productivity and competitiveness through the selection and implementation of the appropriate design tool for their needs. I have a dislike for acronyms and sentences with long words and marketing catch phrases that when paragraphed do not sound like they should be uttered by people in real jobs doing real design work (no offense to marketing departments everywhere), but as drafters, designers, engineers, checkers, project managers, do we really speak like this? I will attempt to keep this book practical and focused on the subject; *why change to 3D, how to change to 3D, and what can you realistically expect to encounter in the process.*

About the Authors

Louis Gary Lamit (lgl@cad-resources.com) is currently a full time instructor and department head at De Anza College in Cupertino, Ca, where he teaches Pro/ENGINEER, Pro/SURFACE, Pro/SHEETMETAL, Pro/NC, Expert Machinist, and Unigraphics NX. Mr. Lamit has worked as a drafter, designer, numerical control (NC) programmer, technical illustrator, and engineer in the automotive, aircraft, and piping industries. He started as a drafter in 1966 in Detroit for the automobile industry, doing tooling, dies, jigs and fixture layout, and detailing at Koltanbar Engineering, Tool Engineering, Time Engineering, and Premier Engineering for Chrysler, Ford, AMC, and Fisher Body.

Mr. Lamit has worked at Remington Arms and Pratt & Whitney Aircraft as a designer, and at Boeing Aircraft and Kollmorgan Optics as an NC programmer and aircraft engineer. He also owns and operates his own consulting firm (CAD-Resources.com - Lamit and Associates), and has been involved with advertising, and patent illustration.

Mr. Lamit received a BS degree from Western Michigan University in 1970 and did Masters' work at Wayne State University and Michigan State University. He has also done graduate work at the University of California at Berkeley and holds an NC programming certificate from Boeing Aircraft.

Since leaving industry, Mr. Lamit has taught at all levels (Melby Junior High School, Warren, Mi.; Carroll County Vocational Technical School, Carrollton, Ga.; Heald Engineering College, San Francisco, Ca.; Cogswell Polytechnical College, San Francisco and Cupertino, Ca.; Mission College, Santa Clara, Ca.; Santa Rosa Junior College, Santa Rosa, Ca.; Northern Kentucky University, Highland Heights, Ky.; and De Anza College, Cupertino, Ca.).

His publications include:

Industrial Model Building, with Engineering Model Associates, Inc. (1981)
    Prentice-Hall
Piping Drafting and Design (1981)
    Prentice-Hall
Piping Drafting and Design Workbook (1981)
    Prentice-Hall
Descriptive Geometry (1983)
    Prentice-Hall
Descriptive Geometry Workbook (1983)
    Prentice-Hall
Pipe Fitting and Piping Handbook (1984)
    Prentice-Hall
Drafting for Electronics (3rd edition, 1998)
    Charles Merrill
Drafting for Electronics Workbook (2nd edition 1992)
    Charles Merrill

CADD (1987)
    Charles Merrill (Macmillan-Prentice-Hall)
Technical Drawing and Design (1994)
    West and ITP/Delmar
Technical Drawing and Design Worksheets and Problem Sheets (1994)
    West and ITP/Delmar
Principles of Engineering Drawing (1994)
    West and ITP/Delmar
Fundamentals of Engineering Graphics and Design (1997)
    West and ITP/Delmar
Engineering Graphics and Design with Graphical Analysis (1997)
    West and ITP/Delmar
Engineering Graphics and Design Worksheets and Problem Sheets (1997)
    West and ITP/Delmar
Basic Pro/ENGINEER in 20 Lessons (1998) (Revision 18)
    PWS (a division of International Thomson Publishing)
Basic Pro/ENGINEER (and PT/Modeler) (1999) (Revision 19 and PT/Modeler)
    PWS (a division of International Thomson Publishing)
Pro/ENGINEER 2000i (1999) (Revision 2000i)
    Brooks/Cole (a division of International Thomson Publishing)
Pro/E 2000i2 (includes Pro/NC and Pro/SHEETMETAL) (2000)
    (Revision 2000i^2) Brooks/Cole
IX Design, CD book, (2001)
    ImpactXoft
Pro/ENGINEER Wildfire (2003) (Revision Wildfire)
    International Thomson Publishing
Introduction to Pro/ENGINEER Wildfire 2.0 (2004) (Revision Wildfire 2.0)
    Schroff Development Corporation (SDC)
Pro/ENGINEER Wildfire 3.0 (2006) (Revision Wildfire 3.0)
    Thomson Engineering (Thomson)

For additional information see: www.cad-resources.com/Other_textbooks

James Gee (jgee@cad-resources.com) is currently a part time instructor at De Anza College, where he teaches Pro/ENGINEER, Pro/MECHANICA, Pro/CABLE, and Pro/MOLD. Mr. Gee graduated from the University of Nevada-Reno with a BSME.

He has worked in the Aerospace industry for Lockheed Missiles and Space Company, Sunnyvale, California; Space Systems/Loral in Palo Alto, California; and currently for BAE Systems in San Jose, California. Mr. Gee has assisted in checking and contributing to a number of CAD articles and the Pro/ENGINEER textbook series with Mr. Lamit.

Acknowledgements

I want to thank the following for the support and materials granted the author:

Gearoid Smyth	PTC
Darcy Parker	PTC
Leslie Minasian	PTC
Mike Campbell	PTC
Larry Fire	PTC
Steve Keith	De Anza College
Thuy Dao Lamit	Lamit and Associates
Dennis Stajic	CADTRAIN-PTC
Max Gilleland	De Anza College
Letha Jeanpierre	De Anza College, Dean of Business and Computer Systems

Industry interviews:

Tim Grady	ACCO Brands Incorporated
Robert Milich	Edwards Life Sciences
Bruce Hillukka	Advance Tool Incorporated
Mark Mcguire	Energy and Power Automation

Wikimedia Foundation: for a description of PLM and PDM, which others can use without permission from this publication.

Download the current Pro/ENGINEER Tryout Edition and view free online tutorials by selecting the following link: http://www.ptc.com/offers/tryout.htm

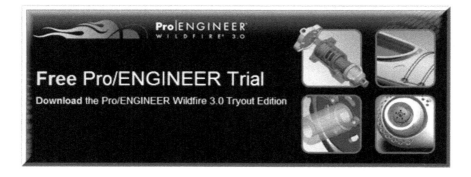

Table of Contents

Introduction (by Gearoid Smyth)

Chapter One   Why change?

Chapter Two   Evaluation through implementation

Chapter Three   Comparing the workflow: Part design
                   (Online tutorial available)

Chapter Four   Comparing the workflow:
                   Assemblies and documentation
                   (Online tutorial available)

Chapter Five   Reusing 2D design data
                   (Online tutorial available)

Chapter Six   Tools for converting 2D to 3D
                   (Online tutorial available)

Chapter Seven   Your workforce

Chapter Eight   Selecting 3D CAD software and hardware

Chapter Nine   What you are missing

Chapter Ten   Success stories

Appendix

Index

Introduction (by Gearoid Smyth)

Every day, it seems there is increasing pressure on discrete manufacturers to get better products to market, faster and cheaper. Companies of all sizes are trying to squeeze every last ounce of productivity out of their people, processes, and tools. Yet the challenges designers and engineers face continue to mount. Are there untapped benefits in moving from a 2D CAD solution to 3D? If you're considering making the switch, what are some things you should keep in mind? This book by Louis Gary Lamit is meant as a guide and tutorial for those seeking information and assistance in the process.

Each year, tens of thousands of CAD users migrate from two-dimensional drafting to the world of 3D solid modeling. Their reasons may vary – manufacturing may demand 3D handoffs, or their companies (or clients) may demand the faster time-to-market that 3D models provide. While millions of product designers and engineers are now using 3D solid modeling CAD tools, there are just as many still using more traditional 2D CAD applications to design and develop their products.

The path from design to manufacturing isn't smooth or direct: it suffers constant dead-ends and reversals. Designs constantly come back, with requests for changes to different parts of the product: the wiring harness, the framework, or some other custom-fabricated component. And while some changes are straightforward, others can have a domino effect that affect the structural integrity of the design and jeopardize the entire schedule.

Beyond delivering cheaper and more affordable products, today's 3D CAD systems help deliver exactly what many customers want. Through the power of parametric 3D models, which can be easily re-used and modified, more and more companies are able to easily deliver unique variants of their products, to meet the specific needs of their customers. From PCs to mountain bikes to fire engines, customers are demanding products designed and built to meet their specific needs. Delivering these products with more traditional methods would be cost-prohibitive, but modern 3D CAD systems allow designers and engineers to edit a few parameters, and automatically create the downstream deliverables for unique variants in minutes, instead of days or weeks.

With all of these benefits, why doesn't everyone use a modern, 3D solid modeling CAD/CAM/CAE system? The reasons certainly vary, but here are some of the key questions that design and engineering management needs answered before taking the plunge.

1) How steep is the learning curve of the 3D application?
2) How effectively can I leverage and re-use my existing 2D data?
3) What is the total cost of ownership of the new 3D tool, including hardware, software and the associated training?
4) How do I know the 3D system will grow as my needs change?

By addressing these concerns here in detail, the engineering manager can compile a short list of 3D software candidates that will represent the best choice in 2D to 3D transition strategy. For those companies that have decided to make the transition, there's never been a better time than now. Today's entry-level 3D CAD systems are well within reach of even smaller companies. And by selecting wisely, product designers, engineers and managers can achieve a solution that's easy to learn, easy to leverage 2D legacy, and easy to afford. Finally, 2D users can enjoy the benefits that thousands of other 2D users are enjoying by 'going 3D'. This book will assist you in this process.

Gearoid Smyth

Chapter One Why change?

- Challenges to change
- Usability and learn-ability of 3D CAD
- What to do about legacy data?
- Cost and Scalability
- Downstream opportunities and capabilities

This book is meant to assist you in your decision to go 3D and to guide you through the process. The best approach is to evaluate 3D CAD software in terms of learning curve, legacy data re-use, total cost including software, training and support, and future scalability.

If you are in the design industry, eventually your company and design department will be faced with the inevitability of transiting your design tool from a 2D CAD to 3D CAD system, either now or in the near future. Instead of "why change?" there is a more important question; with so many of the competition already using 3D CAD, "can I afford to stay with 2D design?" So the question of changing from your existing 2D CAD design tool to a 3D CAD tool becomes; when to start, not if to change. In the end, the real question is not, why change, but how to change. The challenge is to have the transition from a 2D CAD tool to a 3D CAD tool be as seamless as possible, limiting design down-time, expensive missteps, and internal discord.

During the research for this book, there was one quote from a person I interviewed that really struck a chord: *"We simply could not be in business doing all the things we do if 2D were the only option."*

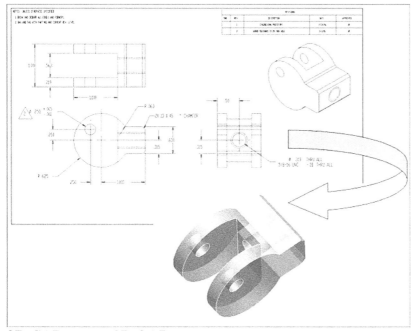

2D CAD versus 3D CAD

All 3D modeling systems may look similar but only on the surface. A wonderful little 3D CAD system may tout itself as being able to compete with the high-end and mid-level CAD systems function to function. But, as I have found out from personal experience, the differences are more than skin deep. In one case, I asked a vendor if their system did manufacturing-NC-machining. The answer was yes! So, I proceeded to ask if they could demo it so I could see the capability. Well, they "really did not have an integrated CAM solution, but you could purchase a 3rd party package and it would run perfectly with their CAD models". Hated to ask, but what was the cost of the external CAM system. Answer - $5,000 (US) for the 2 1/2 D version and up to $20,000 (US) for the multi-axis machining version, such a deal. The CAD software price tag was slightly less than one of the major competitors but by the time I added the external CAM package, it was considerably more, even with the additional cost of the CAM module of the "more expensive system". It was also not a "seamless" database. Changes in the model did not automatically propagate throughout the part-assembly-drawing *and* manufacturing model.

So, the key to success comes down to evaluating how effectively a 3D modeling system addresses the fundamental transition concerns:

- Productivity (usability and learn-ability)
- Legacy (2D designs data)
- Cost (total cost of ownership and cost of use)

By addressing these concerns in detail, the engineering manager can compile a short list of 3D software candidates that will represent the best choices in their 2D to 3D transition strategy. The list will reaffirm the answers to the question; "why change?", since so much of the list expands on the usages of the 3D model database when compared to the traditional 2D design document.

Another factor, scalability, encompasses a variety of areas including:

- Extended design (mechanisms, large assemblies, and industry-specific applications)
- Downstream capabilities and functionality (analysis, mechanism dynamics, animation, rendering, inspection, testing, prototyping, CAM)
- Data management (data vaulting, change management, collaboration, configuration management, BOM management, MRP, ERP). *See Chapter 9.*

Usability and learn-ability Product designers know that designing in 3D solid modeling requires a very different approach from that of 2D CAD design. To ease this transition, the 3D software itself should help wherever possible, it should be simple to use and understand. It should be modern, and familiar, like other desktop applications. It should also make extensive use of automated training and tutorials. Gone are the days of incredibly powerful, but equally complicated 3D CAD systems that only the most advanced users could master.

Microsoft's Windows operating system has been adopted as an engineering standard; more individuals than ever before are interested in using 3D CAD solutions; and no one has a tolerance for spending weeks on end trying to learn overly complex software. The best 3D CAD tools provide powerful capabilities, in a scalable, easy-to-learn and easy-to-use package, that is familiar, and allows designers and engineers to spend their time delivering great products, instead of learning how to use the tools.

Furthermore, the available CAD learning tools must feature comprehensive, built-in tutorials and online training options that allow users to learn various topics at their own pace.

Legacy data With some 3D systems, 2D users are forced to transfer their legacy data to the new system by translating files from one industry-standard format to another. As engineering managers know, this translation introduces substantial data inaccuracy, and often wastes valuable design engineering time. Since this neutral format carries forward a minimum of useful information, the design engineer must spend hours repairing the translated data. Some 3D CAD systems have an available internal, integrated program that will allow the use of 2D data directly as control geometry for 3D features.

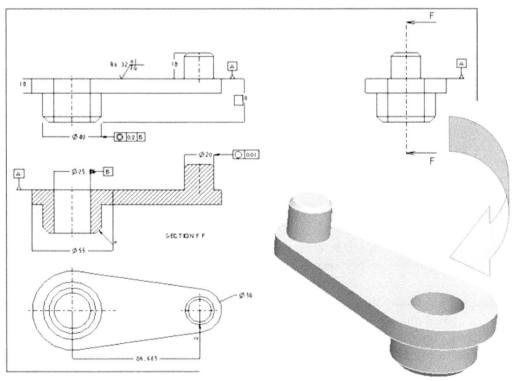

2D CAD legacy transferred to 3D CAD

Cost One of the key factors in upgrading from 2D to 3D CAD software is the fear of the cost to make the move. Most CAD systems cost under $5,000 (US) – typical for a 3D CAD system. But the typical cost of entry-level 3D software can be misleading, since the more important issue is total cost of ownership. Cost of ownership takes into account the cost of software, training and customization, plus the cost and quality of support. Cost will also include the purchase and installation of new hardware (and hardware maintenance). Although hardware is not a hidden cost, training cost can be misleading and more complex than presented by some software manufacturers.

Extended design Entry-level 3D CAD systems are not usually considered appropriate for advanced design work and over time users will quickly encounter the limits of their technologies. For instance, a designer may want to work with more complex geometries, or manage 3D assemblies. When this happens, the reason for the concern – the need for a scalable 3D growth path – becomes clear.

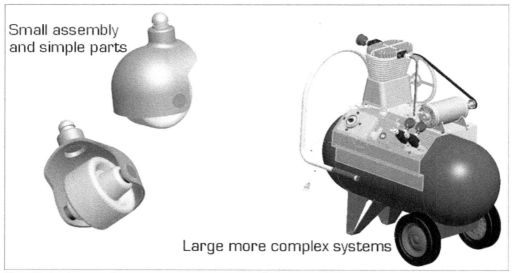

Small assembly and simple parts

Large more complex systems

Small design projects versus more complex assemblies

Downstream capabilities Users of 3D CAD systems can extend the use of the original part database for concurrent engineering processes, which allow downstream deliverables like mold cores and cavities, and NC tool path to be developed, even as the design of the product is still evolving. Through seamlessly integrated, associative CAD, CAM and CAE applications, these deliverables can automatically update, even when changes are made to the design very late in the process, including manufacturing.

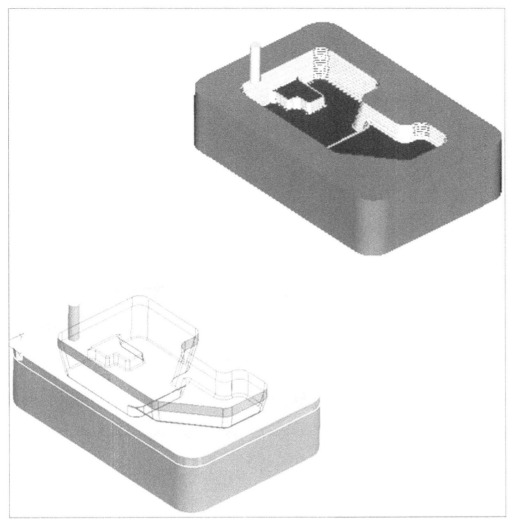

Milling with 3D CAD model

Digital models are now complete, virtual definitions of the final product, and may include all types of product content, such as plastic parts, cast parts, sheet metal components, pipes, cables and even representations of PC boards and other electrical data. By having this complete, virtual definition of the product, designers and engineers are able to confront issues long before any physical tools or products themselves are identified. Basic interference checking, to ensure that components do not clash, is a major benefit. Photorealistic renderings allow marketing materials and product packaging to be produced, well before the physical product even exists. With these capabilities, 3D solid models radically cut the time-to-market.

Benefits of 3D versus the challenges of 2D

Basically, 2D can be cumbersome and difficult to use as a true design tool. The only areas of similarity between these two methodologies are the use of 2D sketching for geometry creation and the availability of a plotted/printed document at the end of the design process.

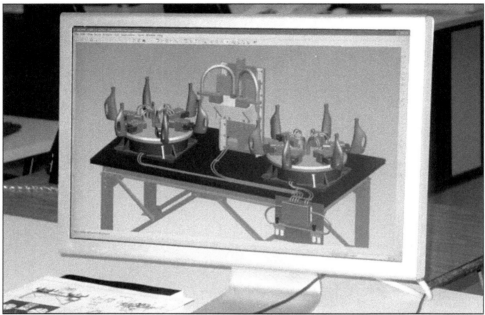

3D Design

Documentation is simpler to extract with a 3D CAD system, though do not be fooled by promises that the "drawings are created automatically". The usefulness and the bi-directional parametric nature of 3D CAD does open up documentation capabilities. Documents are actually intelligent extensions of the total database. With 2D CAD, your documentation package is similar to a book. Nice reading but that's it. The 3D "documentation" package is more analogous to a "DVD"; it may contain drawing files, but also has a complete video-like quality. With an integrated 3D database, changes to documentation propagate to the component and assembly. Changes to the component propagate to the assembly and the drawings, and changes to the assembly propagate to the drawings and to the component.

Regardless of where you are in the process or what form the changes take, they are reflected in the entire documentation, thereby reducing and possibly eliminating mistakes that can spiral out of control when 2D drawings are the control documentation of a product, instead of a product model. Quality control will actually increase exponentially.

Another benefit of moving your design tool to 3D CAD is the use of pictorial views on drawings and for technical illustrations for brochures, manuals, and other sales literature. Animation capabilities also enable you to use the database beyond that of traditional engineering documentation and extend into visual motion presentations.

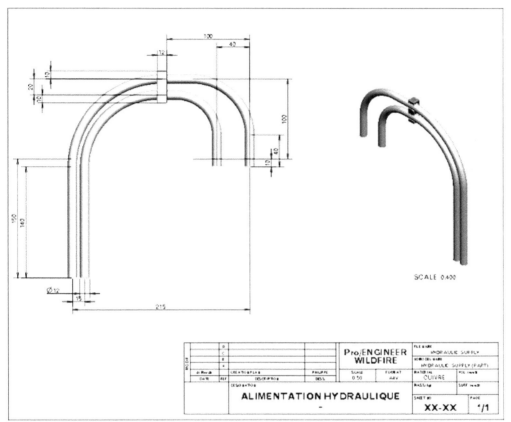

3D CAD detail drawing

Moving from your existing 2D CAD design tool to a complete 3D CAD design tool will affect your time-to-market and what you market. Product line variations will extend your business by propagating new part families since the design variations can be extracted from the same original. New and more varied products born from the same design database will positively affect marketing and sales.

Your move from one design tool to another will not be without hurdles. Going into the process requires that you and your team have every available transition tool, available outside assistance that you trust, and the support from those above you in upper management and below you in your engineering, design, and manufacturing workforce.

This is a complicated and intense journey. Hopefully, this book will assist you in the path to success.

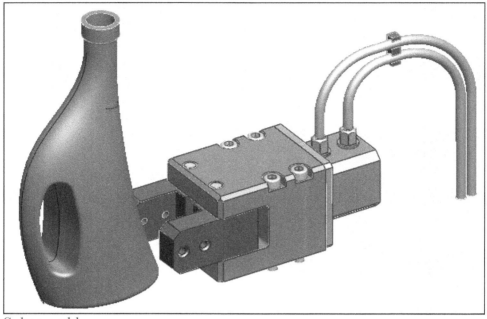

Subassembly

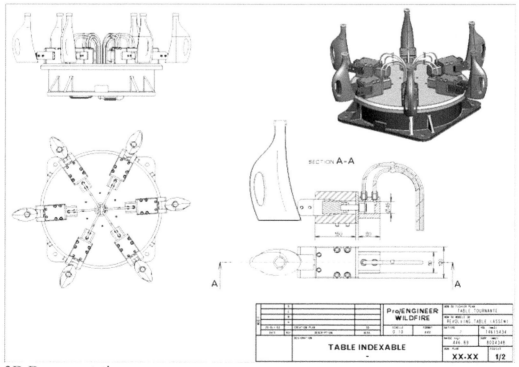

3D Documentation

Chapter Two **Evaluation through implementation**

- Legacy Designs and their reuse
- Company training commitments
- Evaluation of your companies needs
- Setting up a selection team
- Implementing the new system
- The transition process

In the late 1980's I received a call from an "old guy" who owned a mold design and mold making company. I knew he was old because he said so (I am the old guy now). During my career, I had always got along well with "old guys"; tough, seasoned, no-nonsense, experienced, industry-wise individuals who had started or ran their own companies. He asked me about this CAD system called AutoCAD, since he had heard we were offering classes teaching it.

I asked him a series of questions: What does your company design and manufacture; what is your product? How were you doing your designs previously (manual drafting or CAD), and what were you using now? He answered all and then started to comment about the AutoCAD system he had purchased recently. I asked why AutoCAD? He said: "because one of the design team had heard about it and said it was easy to learn and use". I asked how it was going. He said: "Terrible, I am staring at my design team through my office window and they don't seem to be accomplishing anything and haven't since the AutoCAD system was installed! All they do is play! No drawings had been completed, nothing at all!"

I was glad I was in education "working" for the government, where my biggest headache was dealing with a few useless and clueless administrators (and protected by tenure).

I hated to ask him the next question. Why did you buy AutoCAD? It is 2D and you get electronic drawings similar to the ones they were generating by hand. Asked did he look into CAD systems based on price or function? He was kind of quiet. I proceeded to ask if he evaluated a system called CADKEY, which I knew, was 3D and had a mold design module? He said: "No." Since I had been teaching Computervision, a true 3D CAD system (wireframe and surface modeler), and had just instituted the AutoCAD program, I knew the difference between buying and using a "drafting" tool and a "design" tool, and, that not all tools were equal.

The moral of the story is quite simple. Do your homework. Don't listen to some employee who "heard" this or that system was good, easy to learn, and cheap. Do a complete evaluation of your present situation; including your product line now and in the future; your design needs, now and in the future; your facility and design team, now and in the coming years. Do it before you buy. Get the complete story. Find out what CAD system can do what. And most of all; bring the team(s) together and air out what the CAD tool needs to be able to accomplish. My old friend let the tail wag the dog. The dog now had fleas.

This chapter (and the book in general) will guide you in the consideration, evaluation, and implementation of a 3D CAD system with the appropriate capabilities and functionality. And then migrate in a manner that brings your company into the 3D world with a minimum of problems and a maximum of benefits (and will keep you flea-less).

Legacy Design Reuse

What about the other side of legacy designs? You only hear of how much it will take to convert old 2D data. What about how much time and energy is saved when a 3D design can be updated and reused for a similar or new product line, or during redesign and ECO changes? In other words; do not just concern yourself with your "old" legacy data, what about the promising future where your "new" legacy data can be used again and again, used to develop an array of new designs and products. The result is legacy archives, where the database does not ever become obsolete, since it is now "3D" legacy.

Buy what you need with an eye on the future

Don't buy what you won't use - PDM, Analysis? But make sure your choice of CAD systems allows for future downstream additional functionality without having to purchase new software from outside sources. Your system should allow you to expand seamlessly, within the 3D CAD product functionality without learning a completely new interface and new software. Purchasing and implementing a 3D CAD system is a big commitment. In a way it is like getting married. You're in it for the long haul. Plan, commit, and work on making it a success. Just as in a marriage, it's a lot easier to make it work if you selected the right partner. And it's very painful and expensive if you chose unwisely. Implementing and then disengaging from a design tool is no easy matter.

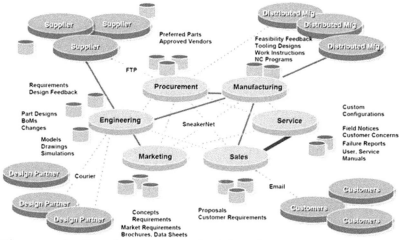

The complexity of modern product development

Internal training commitment

Training is not just a one time one week $1000 (US), buy it, do it, and it's over situation. It is a long term continuously evolving requirement that needs to be planned for, and funded adequately. You need to explain to your design team how 3D CAD skills will benefit their personal career and secure their place in the company. A high-level quality 3D CAD system will draw them in and keep their interest. Buying and implementing a cheap, non-mainstream CAD tool [GonzoCAD for $295.00 (US)] will not entice them to change or come up to speed using that tool. All 3D CAD systems have basic packages for about the same price, $3000-$5000 (US). Functionality, for that same dollar, is very different.

Show your existing personnel how this new tool benefits their job by increasing job skills and their design efficiency. Evaluate a system based on whether it will meet your needs but will also be recognizable to your existing staff. Make sure there is local talent available. The new 3D CAD system should have local educational possibilities. Be willing to work with local educational facilities to get the CAD tool taught. Get real assistance not just a quickie from some CAD "expert" who "guns and runs" and in the process offends your existing engineers and designers. You need long-term assistance that really helps your staff come up to speed and keep them "in training".

3D CAD training does not have to be cost prohibitive

Steps in the process

In general, follow the steps of evaluation, selection, implementation and transition. Every company will be different and have varied needs and requirements depending on their product sophistication and industry specific parameters. Some elements of this discussion will apply to all enterprises and some only to a select few that fall within the stated parameters. Regardless, develop your own benchmarks, requirements, and lists for completing this process. Remember, every up front hour spent laying out a logical and comprehensive plan based on your specific situation will save hundreds of hours when the proper CAD system is selected and implemented.

- **Evaluation** – Create a list of evaluation questions that need to be answered by your design team and the numerous CAD software company representatives.
- **Selection** – Decide on the CAD system that best fits your company's present needs and with an eye on the future.
- **Implementation** – Develop flexible but tough target dates for the installation, set-up, rollout and maintenance of the CAD system.
- **Transition** – Introduction, familiarization, training, and daily use of the CAD system.

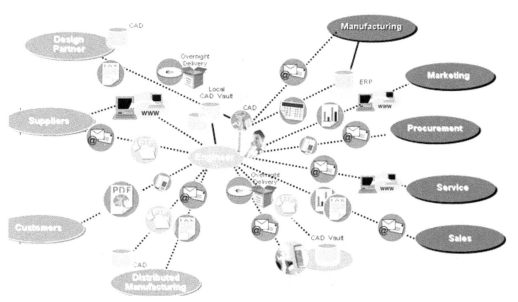

What a 3D CAD system can do

Evaluation

Have your design team evaluate your present product line and identify the possible advantages of installing a 3D CAD system. Contact each CAD software company and setup meetings for demonstrations and benchmarks. Make sure that you have a typical design scenario available to use for the benchmark. Some companies may choose to have the software companies perform the benchmark testing. You may choose not to give the CAD software company pre-access to the material for them to "can" a demo and skew the benchmark in their favor.

Ask that the 3D CAD software company representation not be just a sales representative. Make sure they provide an engineer who knows their system and the standards of your industry not just a whiz-bang CAD demo person with no real world experience. If you are unsure about who they send have one of your seasoned and "tougher" engineers or designers at the meeting who will ask the tough pointed questions (you know who I mean). Do not be afraid of "offending" your contacts. Remember this is your company and design team and it's your neck if it fails. That hotshot CAD rep could possibly be off to another job by the time your company makes a decision.

Also consider and evaluate the hardware that 3D CAD software requires. You do not want to hear from your design team that the 3D CAD tool is running too slow or crashes often because their computers are too slow or do not have enough memory.

Create a budget for the software and hardware and include basic training and long-term support (maintenance).

How much time do your teams spend searching for information? How do you promote reuse?

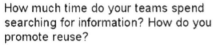

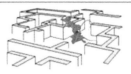

How quickly can you respond when sales changes customer requirements (ECOs)?

How is Bill of Material (BOM) information exchanged between Design and Production?

What is the significance of achieving and maintaining regulatory compliance?

How do you insure remote sites, suppliers and customers are working from the correct versions of information?

What are your company's opportunities for improvement?

Start the evaluation stage by answering a few basic questions:

- Is the present system (manual or CAD) adequate for our present and future?
- Will a 3D CAD system allow reuse of design data in a way that will improve product changes and iterations?
- Does our business model and product sophistication warrant a change?
- Are the development and design times required to bring our product to the marketplace improvable through a better design tool?
- Would changing to a 3D CAD system increase our companies market share and possibly allow us to expand into parallel fields and open up new product lines?
- Are our competitors using 3D CAD?
- Does the new 3D CAD system have a seamless connection to the Internet? This is important for connecting multiple design sites and taking advantage of international business opportunities and resources.
- Do I need a 3D CAD system that will interface with my customers system?
- What do our vendors use as their design tool?
- Who does our manufacturing and where? Will we ever manufacture in-house?
- Does manufacturing have the ability to use the 3D models from the system we choose? Will the 3D database be compatible with existing MRP (Material Requirements Planning) and ERP (Enterprise Resource Planning) systems?
- What is the complexity of our product – 10-10,000 parts?
- Do we need large assembly capabilities?
- Can I identify the primary use of the CAD system based on what our product requires: mold design, sheet metal, machined parts, castings, forgings, etc?
- What functionality is essential for the new 3D CAD system to include?
- In a logical analysis, what capabilities and software options are needed?
- Will we need to customize the new software or configure it to our company's specific requirements or standards? If so, what will be the added benefits? How long will this take? Who will maintain the customization?
- Estimate what percentage of the legacy 2D design data should be converted? Who is going to convert the data? How long will this take? When should it be performed?
- How will the 3D CAD tool be implemented?
- Is the 3D CAD software easy to use?
- How much money is available for training?
- Will our products benefit from the availability of PDM (Product Data Management) or PLM (Product Lifecycle Management)?
- What will the new 3D CAD system cost per station?
- Will a sophisticated server environment be required? Will our present system and network be adequate?
- Are our current computers, operating systems, and monitors adequate? Do they need to be up-dated or replaced?
- Is there complete commitment by company upper management for new CAD equipment and for CAD software support and staff?
- Do the costs of change outweigh the benefits?

Selection

When the time comes for selection, you and your evaluation team should have satisfactorily answered most if not all of the questions posed in the evaluation stage. Every major system issue should have been vetted by this time. An agreement by the whole team to push forward and select a CAD system is at hand.

What if there is discord in the team and in the selection? These problems need to be ironed out and an agreement made to support the decision by "all" involved. There may be varied opinions and preferences that will always surface. But, it is very important that the team and the organization be committed to a united selection. Without a consensus, there will be individuals who cause discord in the ranks that can destroy a well thought out plan for implementation and transition.

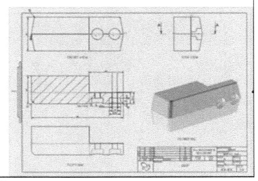

- Create 2D drawings according to international standards
  - including ASME, ISO, JIS and DIN
- Automatically create an associative BOM and balloon notes
- Use templates to Automate the creation of drawings

- Benefits:
  - Accurate BOM on Drawings
  - Compliance with company and international standards
  - Rapid drawing creation

What can 3D CAD bring to the table?

- Establish a team composed of members from your design workforce at each level and the system manager(s). This team of (4-5) individuals will be tasked to monitor the strategy of selection and to submit progress reports so that the process moves forward evenly and efficiently. Ensure that the appropriate department heads within your company receive these reports.
- Be sure to keep affected departments in the decision loop. Seek their input and address their concerns. Someone left out could adversely affect the best laid out plans. But at the same time, you don't want to bog down the process.
- Write specific technical requirements and procedures for establishing the 3D CAD system as the primary design tool.
- Rate each requirement as to its overall importance and function in the design process.

- Get an agreement on the budget from upper management. Include software, hardware, transition, training and maintenance costs.
- Estimate the effects of downtime on your product development during the implementation and transition of the new design tool.
- Get the 3D CAD companies to loan your facility an appropriate computer and loaded software. Make sure they provide one that would be comparable to that which you have or plan to purchase within your hardware budget. If you have a very small enterprise they may not wish to do this. Make it a requirement for you to consider their product.
- Test the 3D CAD systems being considered as to their capabilities and functionality as they would relate to specific aspects of your unique enterprise.
- Evaluate the ability of each 3D CAD system to perform your benchmark requirements and compare the results.
- Can collaboration between various design sites be seamlessly employed? Make sure your benchmarks include testing this capability.
- Does the 3D CAD system have integrated and easy to use Web-based communication tools for distributing data with vendors and customers?
- Select the 3D CAD system based on a team assessment of the potential success of its implementation.

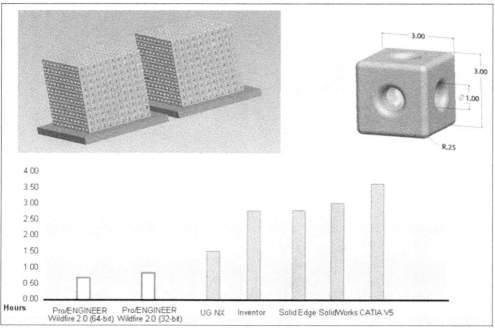

Assembly benchmark

Implementation

Each company or department will determine the implementation sequence differently since the product line and the complexity of the enterprise itself drive it. Introducing a 3D CAD system will normally begin with using it in the 3D design of the product with documentation (details and drawings) following. In many cases, the documentation area may lag behind if due dates for the product are immediate and pressing.

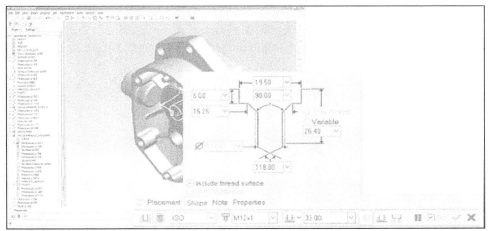

What is the interface like to use?

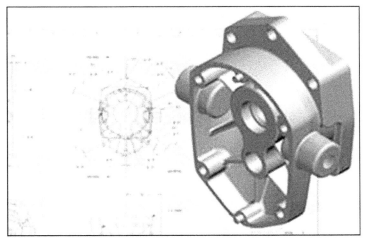

What can be expected in functionality?

- Institute realistic target dates for implementation of a 3D CAD system, and its specific requirements.
- Put in place a mechanism for review and enforcement of timelines.
- Create a set of design standards for 3D CAD functionality as they relate to your product, as well as guidelines for compliance to those new standards.

- Introduce 3D modeling (part and assembly) and documentation (detail and assembly drawings). Then, add a product-determined set of supplemental modules such as CAM (Computer Aided Manufacturing) and NC (Numerical Control).
- Make available a set of hard and soft versions of CAD documentation for easy retrieval by engineers, designers, and drafters.
- Components (legacy or otherwise) that are internally standard and multiple-use (library) parts should be modeled in the early stages as per your company standards and practices.
- Establish (or hire) a CAD system administrator who knows the design variables of your company, has had sufficient training in the use of the new 3D CAD package, and who has the people-skills necessary to communicate and lead the transition, training, and support.
- Create and hold to well-designed performance reviews of 3D CAD users in your company.
- Exchange ideas and concerns bi-directionally between the system manager and the complete design team.
- Is the existing flow of design through manufacturing organized correctly for a 3D CAD system? What changes can be made to take advantage of 3D CAD data? Establish a flow chart and see where changes in the flow of design data can be maximized and reused interactively.
- Consider new ECO and paper (perhaps paperless) approval procedures and processes that may have been cumbersome in the past. Establish metrics for a new and efficient use of 3D design data.

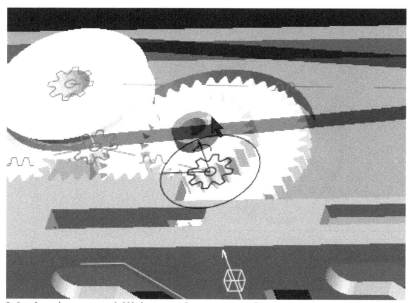

Mechanism capabilities are important for some industries

Transition

The size and makeup of your design department will drive the transition stage. Many if not most companies still have engineers, designers, and drafters.

By the way, where did the checkers go to? After interviewing a number of CAD managers and design team leaders for this book, I concluded that upper management thought that the checkers were not necessary after they had installed a 2D CAD system years ago. Every one of these professionals thought that the checkers should stay, but were overridden by upper management. Hey, upper management – they know everything, right? So it seems that the typical situation is where the engineers have been called upon to do design through documentation when possible (and check). Is this what you really want your (in some cases – highly paid) engineers to be doing? Designers still fill the gaps between engineers and manufacturing. And drafters, where employed, are assigned the tasks related to detailing, ECO's, and document control.

A sophisticated 3D CAD system should help reduce (not eliminate) many of the mistakes, problems, and failures typical of 2D based designs. And though the 3D CAD system may not completely replace that crusty checker, it can provide checks and balances in your entire design through documentation through manufacturing process.

With this scenario, the logical place to start is getting the primary design engineers and designers trained to use the 3D CAD and working in teams to insure that the new 3D CAD product design tool is integrated into their (daily) workflow.

If your workforce sees this new 3D CAD system as an opportunity, that they are part of a team in its mastery, you will have fewer laggards and complainers amongst them. I always ask my students; "Do you want design or whine? I looked in the last Sunday paper employment advertisements and there was not one job for whiners this week".

Develop a splash screen that opens when a user logs on. One CAD manager I talked to said that they have (on this splash screen) a variety of tools and links that would assist the user: Help, CBT (Computer Based Training) training module(s) (they used a product from CADTRAIN), ASME Y14 documentation, company standards and resources, phone numbers, etc.

Although every company must prepare its own appropriate list of transition steps, the following will start you off with a basic framework:

- Most CAD software companies have user interface guides (sheets) that describe their interface and use of the system, mouse control, and hot keys. These sheets will speed your users understanding of the 3D CAD interface and provide a quick reference for overcoming (sometimes) simple but frustrating command and system required input.
- Establish a short period (session) each day for team members to meet and discuss successes and frustrations. Plan on a show-and-tell time in these sessions. Show simple 3D models using a few new features each day. The same will be applied to the assembly and detailing functions. Have a different employee demonstrate something that they discovered or mastered. You can implement this just before lunch 11:30-12:00 noon. Beware, 3D CAD is seductive and frankly amazing in its functionality, they may want to have working lunches!
- Convert only the necessary existing (2D) design legacy data. Put in place procedures and a logical timeline for specific conversions to be completed.
- Establish downstream interfaces with manufacturing and other departments that will use the data created with the new 3D system. There will be more use of the original design data now that a 3D model is generated.
- Integrating some sort of data management tools for controlling the product development and documentation may be important even at this early stage. A full PDM integration will seldom be implemented at this point, but if possible, it should be planned on, and if possible put into use at a very simple level, the sooner the better.
- Train, train, train.

See your 3D CAD training as someone would an exercise program. To stay in-shape you must exercise 20-40 minutes daily (please do not let my spouse see that I wrote this since it would be used against me). Regardless, establish an exercise-training program for your design team members and oversee it as would a personal trainer. After the initial shock of the effort required, it will be second nature for your workforce. Also, lead the way. Add a component of 3D CAD training to your already busy schedule. It will pay off.

Chapter Three
Comparing the workflow: Part design

- Design using 2D CAD
- Design process using 3D CAD
- Capturing design intent
- Design changes in 3D CAD
- Manufacturing using 3D CAD

This chapter will compare a typical 2D part design and detailing with a 3D modeling and detailing workflow. The 3D model will then be used in the manufacturing (milling) sequence.

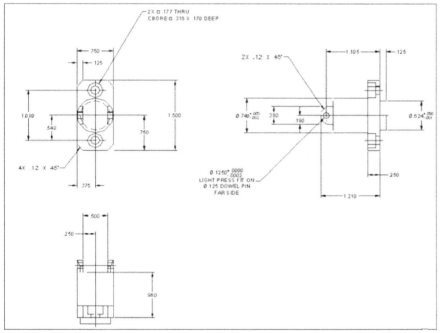

Part design in 2D

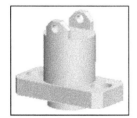 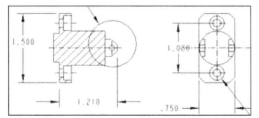 

Part design in 3D

Note: An online tutorial is available at www.cad-resources.com that will guide you through the steps required to create a 3D CAD parametric model. The tutorial assumes that you have not used Pro/ENGINEER Wildfire 3.0 previously. For online files, click www.cad-resources.com > **PTC Partnership** > **Chapter 3 ...** > download and open the zipped files.

2D CAD (left pages)

In 2D CAD, drawing a detailed part or assembly in orthographically projected views can require considerable time and effort, with the end result being a paper document. A 2D drawing also requires effort to interrupt during downstream activities i.e. mold design or a set of NC sequences. Drawing view construction requires repeating simple entity commands and adds considerable time and expense to the design process, one that is subject to a variety of incorrect representations, especially for complex components and assemblies. Changes to the 2D drawings are also time-consuming and repetitious since the modifications must be made on every sheet the feature appears on. The greater the complexity of the component, let alone the related assembly, will add hours to the project and multiply the possibility of mistakes. Though on the surface 3D CAD modeling seems more complex, it is actually less so. Parametric 3D CAD designs are bi-directionally associative. Changes made on the component, the assembly, or the drawing, propagate throughout the documentation package and related models, including the milling sequence if the CAM package is not external to the CAD system.

In the design and engineering world, projects are continually being modified, even at the manufacturing stage. 2D drawings require complex modifications and redraws throughout the lifecycle of the product. Since the 2D drawing process requires considerable work, the use of added, extra, or "unnecessary" views, etc. is common. In 3D design the addition of a view, a detail, a breakout, a rotated view or section is considerably simpler since the view itself does not need "drawing" it needs modification and cleanup at most.

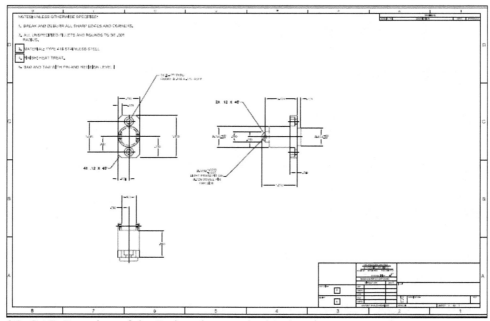

2D CAD Drawing of the Pedestal created with drawing commands

22

3D CAD (right pages)

In many cases, the result of the design process will still be detail or assembly drawings. But, with 2D CAD, that is the end of its useful life. In 3D CAD, the documentation package is only one aspect of the project. The modeled part, describing its geometry, tolerances, material, etc. can be utilized in assemblies, analysis, manufacturing, and even for pictorial rendering for advertising and presentation. After the project is modeled, views for a drawing can be automatically displayed and scaled, including orthographic projections, sections, rotated, detail, or pictorial views, and exploded assembly views.

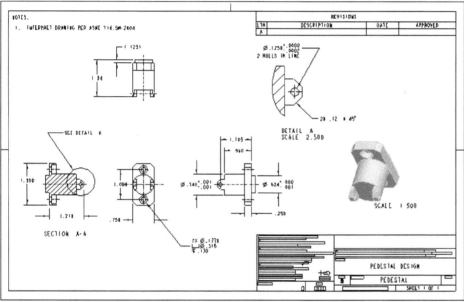

3D CAD Pedestal drawing generated from the 3D model

For drawings, dimensioning starts with the display of feature dimensions established during the 3D design-modeling phase. Displayed dimensions can be placed and moved as needed. Reference dimensions can be created directly on the drawing. The driving dimensions (design dimensions) are normally displayed and then cleaned up so that the drawing conforms to the appropriate standards. ECO's and other design modifications are applied to the 3D model, not the drawing. The drawing will update appropriately as per the modifications, including bills of materials, tooling, manufacturing sequences (milling), and related molds and fixtures. The review of updated drawings is usually necessary since some edits may be required. During every stage and process, the 3D model is used in the design-through-manufacturing sequence. Changes made to the 3D model will propagate through out the entire sequence. ***Change the 3D model and all other processes change/update*** *(assembly, detail drawing, assembly drawing, NC sequence, etc.)* This associativity enables the 3D CAD process to minimize errors and reduce the time-to-market.

2D CAD Front view is drawn by connecting individual lines

In 2D CAD, the design is transformed through a series of sketches and then views of project geometry.

In most modern 3D CAD systems, you create a part from a conceptual sketch through solid feature-based modeling, as well as build and modify parts through direct and intuitive graphical manipulation.

A feature is the smallest building block in a 3D CAD part model. If you build your models with simple features, your parts become more flexible. There are many kinds of features that you can create on a part. There are solid features, datum features, curves, and surface features, and features specific to applications. Part modeling refers to the creation of solid features and some user-defined features. Some features add material and some remove material. The most basic way to add material is through an extrusion. The most basic way to remove material is through a cut.

After you select the desired references on your model, sketch the section geometry. The system adds dimensions and constraints automatically as you create the section. Redefine the dimensioning scheme and constraints, as needed. You can modify the dimensioning scheme created by the Sketcher by adding your own dimensions and constraints.

You cannot explicitly delete any system dimensions. As you add dimensions and constraints, the system automatically deletes system (weak) dimensions and constraints that are no longer necessary. If you want to keep the system dimensions and constraints, strengthen them before exiting the Sketcher.

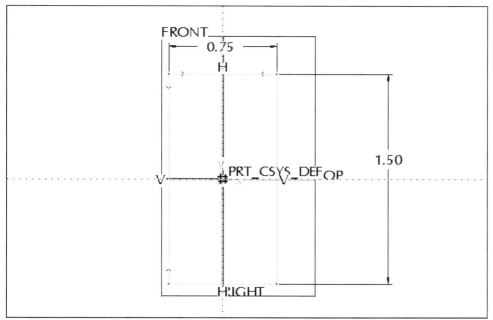

3D CAD The first feature is created by sketching the geometry similar to that used for 2D drawing construction of a front view

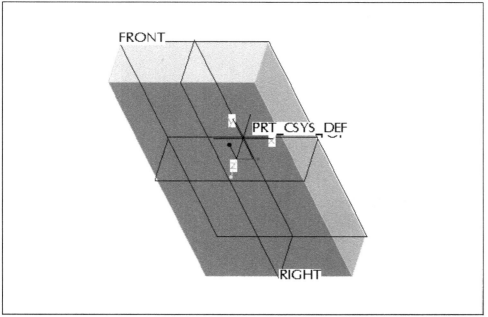

3D CAD Base feature is extruded to the design thickness. A base feature is the first solid feature of a 3D model and will be the parent of subsequent features.

In 2D CAD, the designer creates individual views similar to pencil and paper methods used prior to the introduction of CAD.

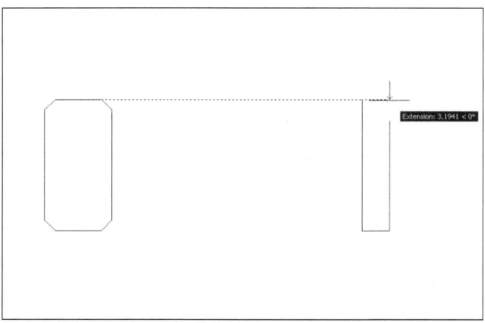

2D CAD Using an orthogonal tool, the drafter creates the right side view

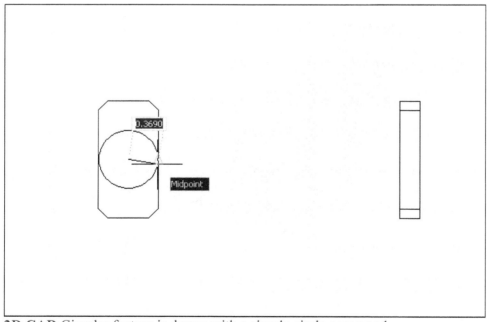

2D CAD Circular feature is drawn with a simple circle command

The next feature will be the short cylinder on the bottom of the part. You will use the same sketching plane and references as in the first solid feature.

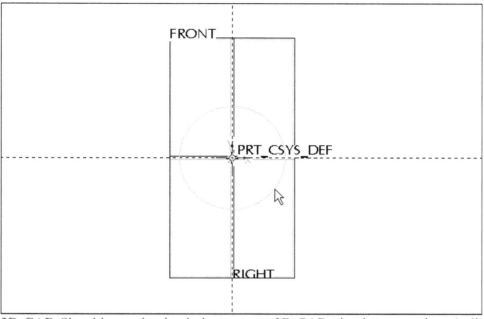

3D CAD Sketching a simple circle creates a 3D CAD circular protrusion. A dimension value is assigned, which you then modify to the design size.

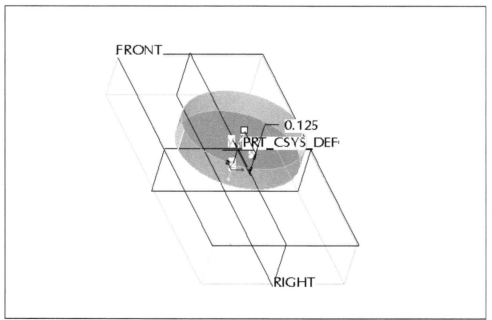

3D CAD Circle is extruded to the design size by modifying the depth value

Parametric feature-based modeling

A **parametric feature based modeler** is a CAD/constructive solid geometry modeling system that enables a user to refer to features instead of the underlying geometry. In 3D CAD, parametric modeling is the term used to describe the capturing of design operations as they take place, as well as future modifications and editing of the design. The order of the design operations is significant. Geometry is associative.

For an example of associativity of features; suppose a designer specifies that two surfaces be parallel, such that surface two is parallel to surface one. Therefore, if surface one moves, surface two moves along with surface one to maintain the specified design relationship. Surface two is a **child** of surface one in this example. Parametric modeling software allows the designer to **reorder** the steps in the object's creation.

For 2D CAD, the completion of the drawing marks the end of its functional use in the design process. Any changes have to be made to the drawing and noted as ECO's, etc. to control the documentation and change flow. If there are multiple sheets, then each view displaying the design change has to be made. With 3D CAD, the change to the model will propagate through all views and all sheets of the project.

In the Pedestal component modeled in 3D, the slot on the top of the part was modeled as a cut referencing the cylindrical extrusion's diameter. If the diameter of the parent changes, the child- cut feature will follow the modified geometry. The cylindrical extrusion is the parent of the slot cut.

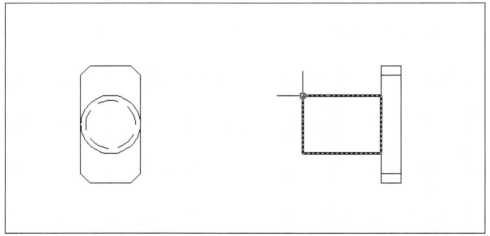

2D CAD The designer continues working between views to construct the 2D representation of the part. Each point, line, or other entity has to be created in one view and then added as a projection to the adjacent view.

28

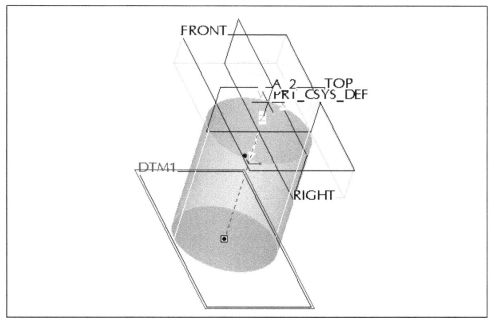

3D CAD The height of the cylindrical feature is extended to a construction datum plane

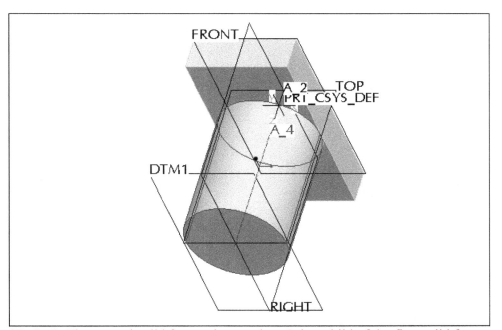

3D CAD The second solid feature is complete. It is a child of the first solid feature.

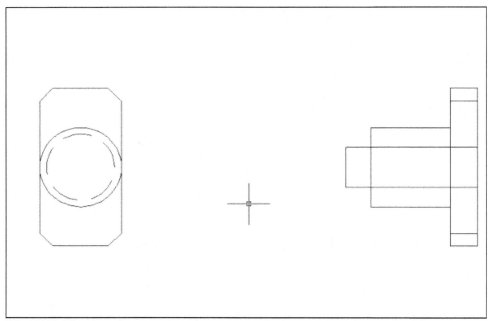

2D CAD The drawing construction process is time consuming and subject to error

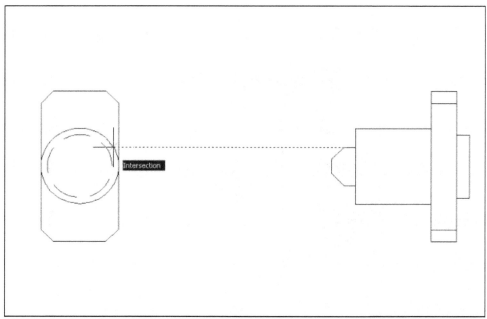

2D CAD The slot on the top of the part is drawn using orthographic projection in the same way pencil and paper were used previously

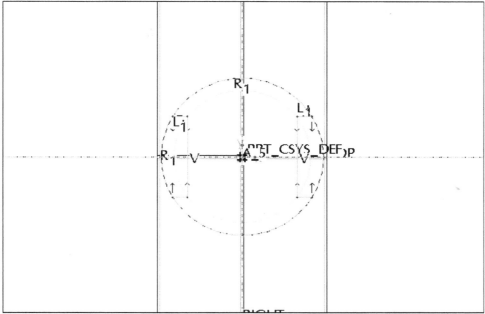

3D CAD The slots' geometry is sketched

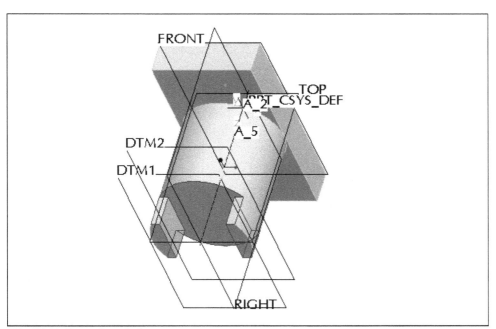

3D CAD Completed slot

Capturing Design Intent

A valuable characteristic of any design tool is its ability to render the design and at the same time capture its *intent*. Parametric methods depend on the sequence of operations used to construct the design. The software maintains a *history of changes* the designer makes to specific parameters. The point of capturing this history is to keep track of operations that depend on each other. Whenever the 3D CAD tool is told to change a specific dimension, it can update all operations that are referenced to that dimension.

For example, a circle representing a bolthole circle may be constructed so that it is always concentric to a circular slot. If the slot moves, so does the bolt-circle. Parameters are usually displayed in terms of dimensions or labels and serve as the mechanism by which geometry is changed.

The designer can change parameters manually by changing a dimension or can reference them to a variable in an equation (**relation**) that is solved either by the modeling program itself or by external programs such as spreadsheets.

Features can also store non-geometrical information. This information can be used in activities such as drafting, numerical control (NC), finite-element analysis (FEA), and kinematics analysis. Capturing design intent is based on incorporating engineering knowledge into a model by establishing and preserving certain geometric relationships. The wall thickness of a pressure vessel, for example, should be proportional to its surface area and should remain so, even as its size changes.

The chamfer feature is added to the first rectangular extrusion. The chamfer is the child of the extrusion. Because solid modeling is a cumulative process, certain features must, by necessity, precede others. Those that follow must rely on previously defined features for dimensional and geometric references.

The relationships between features and those that reference them are termed *parent-child relationships.* Because children reference parents, parent features can exist without children, but children cannot exist without their parents. Since chamfers and rounds are smaller less important features, they are normally added near the end of the modeling process.

The design intent of the part will determine the correct sequence of modeling features. Since chamfers and rounds are not usually integral to the overall design, they can be left for the final stages of construction.

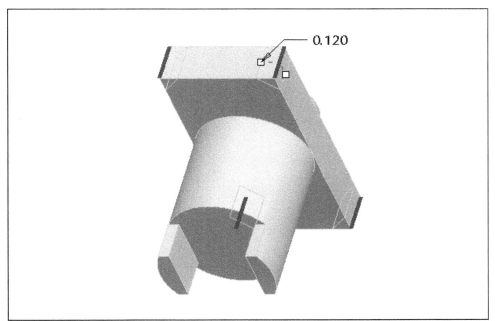

3D CAD The chamfers are added to the base as pick and place features. Select the edges to be chamfered, choose the command, and modify the value by moving the drag handles or typing the design value.

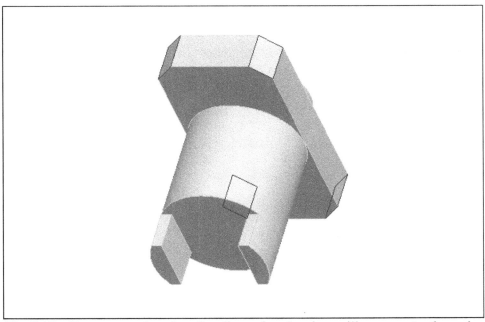

3D CAD Changes to the chamfer dimensional value will propagate through all edge selections

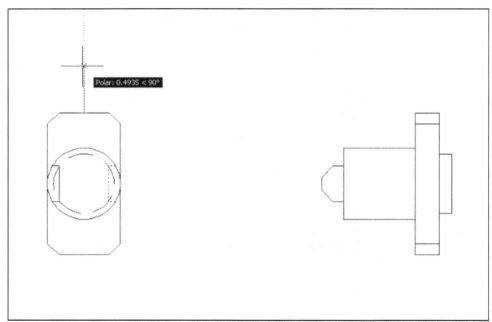

2D CAD The designer uses orthogonal snap to create projected views

As can be seen, the process of constructing drawings is tedious and time-consuming and when completed, the views and documentation are subject to the abilities of the person drawing and annotating them.

In 3D CAD, the modeler (drafter, designer, or engineer) must of course be proficient in the CAD program, understand the design intent to be communicated and understand the product requirements. But, when the model is completed, there will be considerably fewer problems with misunderstanding the design requirements since the model will form the basis of all other down stream steps.

For an example, in the Pedestal part we have two identical holes. In the 3D CAD example, the first hole is created as a pick and place feature. The feature is then selected and then mirrored (the holes could also have been patterned). Modification to a copy affects all members. This helps capture design intent by preserving the duplicate geometry of pattern members.

If the size of the hole changes on a project created in 2D CAD, the designer must edit all drawings where the hole is displayed. The correction-ECO must be made to the drawing geometry and to the features' note. In 3D CAD, the note will update as per the change to the model.

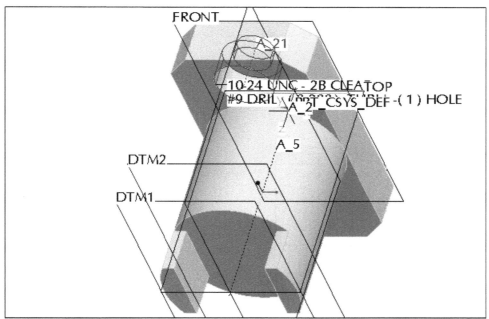

3D CAD Initiating the Hole Tool, picking the placement plane, and selecting the dimensioning references create the standard hole. The type, depth, and other options are keyed into a dialog box.

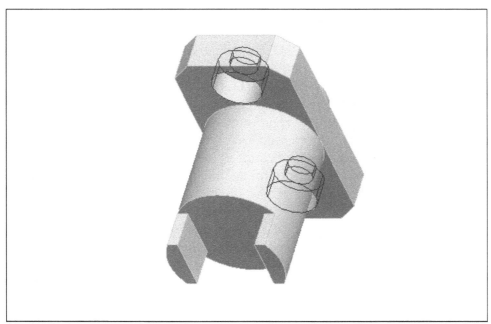

3D CAD A mirror tool is used to complete the two holes. Changes to the initial hole will update the second mirrored hole.

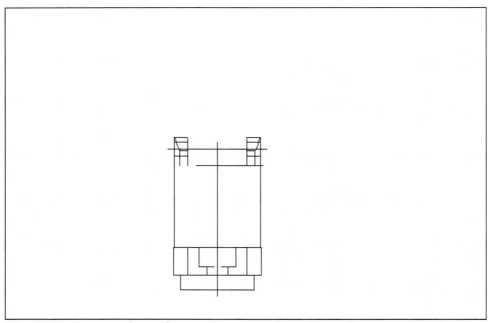

2D CAD Construction of two views requires the designer to visualize each view individually and as a projection in order to represent the object correctly

Creating views of complex objects requires a skilled drafter that can visualize 3D design and represent it on paper/screen as 2D using orthographic views. 2D drawings can easily misrepresent the actual edges and views of complex shapes. Sections drawn in 2D, though used to clarify internal features can lead to even more omissions that result in the engineers or designers original intent being mistaken and result in costly ECO's and scrap parts.

3D CAD models are the parents of each view on a drawing; therefore the internal features are always properly represented on common orthographic views and when internal features sections' are utilized.

Pictorial views require even more effort so as to accurately describe the object being drawn. 3D CAD systems eliminate this process and the possibility of improper and incorrect presentation of the object since the pictorial is derived from the model itself.

So, if you were using a 3D CAD system, the part would have been completed after modeling eight features. The new model can now be leveraged into multiple drawing sheets, unlimited amount of descriptive views for detailing the part, and a database used directly for NC machining. These are but a few of the multiple uses of the 3D CAD model.

On the other hand, if you used a 2D CAD system, you would have a sheet of paper with a representation of the part, and?

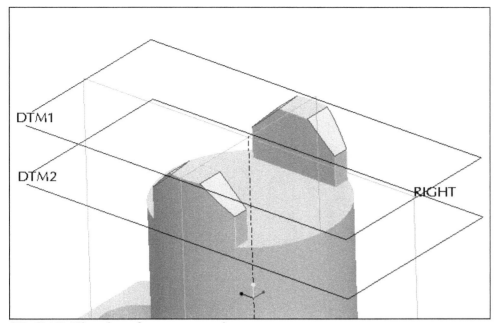

3D CAD The chamfers are complete

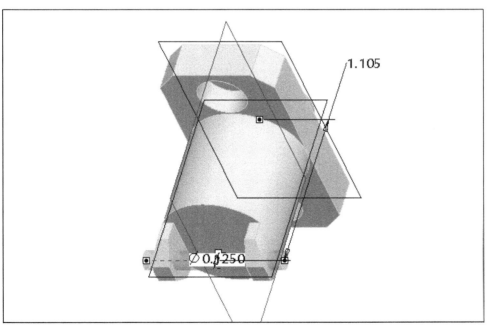

3D CAD The 3D model is complete. A drawing will be generated using the model geometry and the dimensions you used for the construction of each feature.

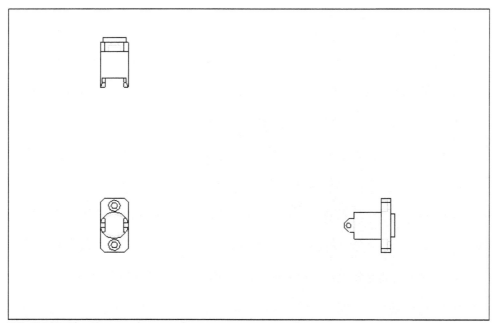

2D CAD The three primary views are now complete

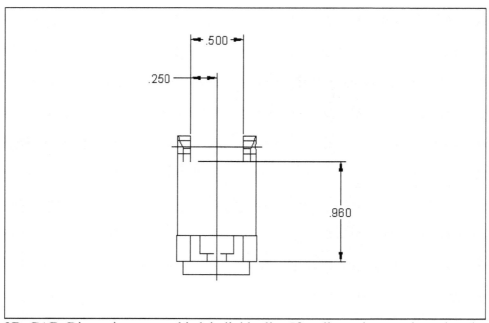

2D CAD Dimensions are added individually. If a dimension needs to be changed to another view, the designer must delete the dimension and recreate it on the view where it better describes the geometry.

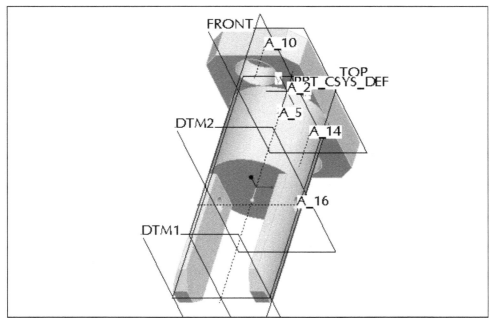

3D CAD Modifications to the geometry will result in immediate model updates after the model is regenerated. The UNDO capability allows the designer the flexibility of undoing and redoing features to try design variations.

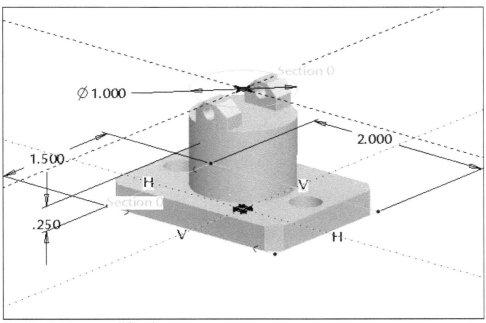

3D CAD More modifications

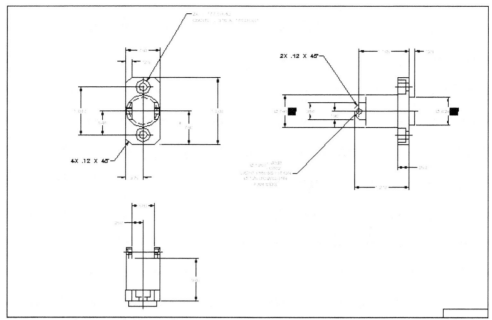

2D CAD Views and dimensions

Please understand that one of the myths of 3D CAD is "your drawing will be automatically created". Not really. There will be considerable time required to position and edit views, dimensions, notes, etc., according to ASME standards.

What you do gain from 3D design is the functionality of ECO's automatically updating the drawing when the original design changes. The use of the original database as the master (unified database) for documentation, manufacturing, analysis, and rapid prototyping are just a few of the many benefits of 3D CAD design over 2D drawing.

You can create drawings of all parametric 3D design models. All model views in the drawing are *associative:* if you change a dimensional value in one view, other drawing views update accordingly. Moreover, drawings are associated with their parent models.

Any dimensional changes made to a drawing are automatically reflected in the model. The original design intent of the engineer will always be imbedded in the 3D model, which in turn drives the drawing and other operations in the process: i.e. assembly, analysis, manufacturing, etc.

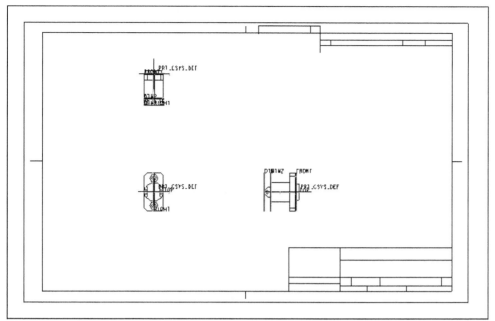

3D CAD Displayed 3-view drawing. Views are generated from the model they are not constructed. The hidden lines of internal features are visually correct.

Any changes made to the model (e.g., addition of features, deletion of features, dimensional changes, and so on) in part, sheet metal, assembly, or manufacturing modes are also automatically reflected in their corresponding drawings. The drawing will always be visually correct as far as views (hidden lines, etc.), sections will be correct, and specialized views (detail, auxiliary, revolved, etc.) can be inserted, not drawn, as required.

"If" the designer used the design dimensions in the original construction of related features, almost all of the required dimensions will display automatically on the drawing, as will centerlines, holes notes, etc.

Hole notes will require editing to bring in line with standards and/or company practices. Drawing notes will require input or use of an existing note file.

41

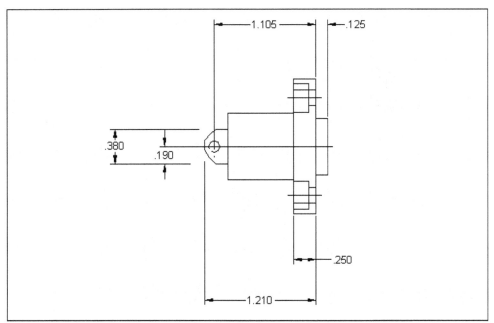

2D CAD Dimensions are placed and edited as required

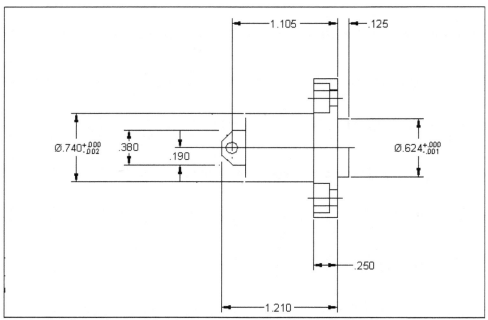

2D CAD Dimensions with tolerances are also added individually

42

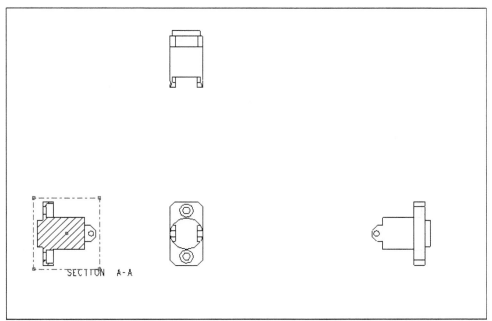

3D CAD Placing a new view, selecting options to convert the view to a section view, which was generated on the model, creates 3D CAD section views

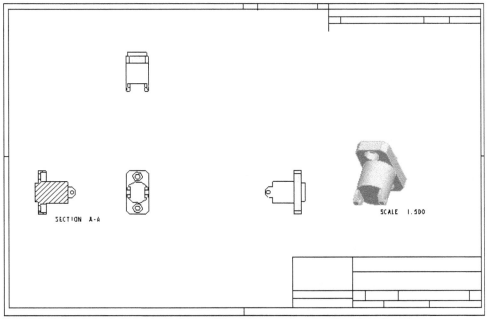

3D CAD A pictorial shaded view of the 3D part is added to the drawing

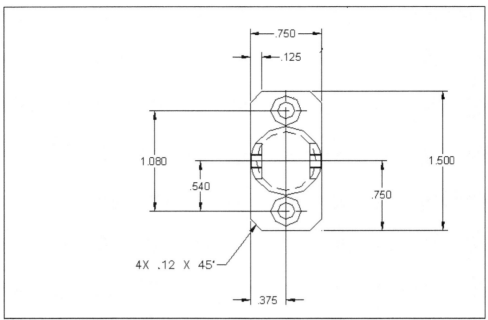

2D CAD Dimensions are added to each view as needed

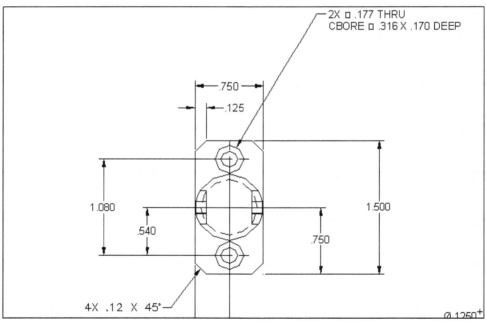

2D CAD Notes are created. If the designer inputs the wrong information it will be transferred to manufacturing. Changes to the design require that the geometry describing the hole and its note both be edited.

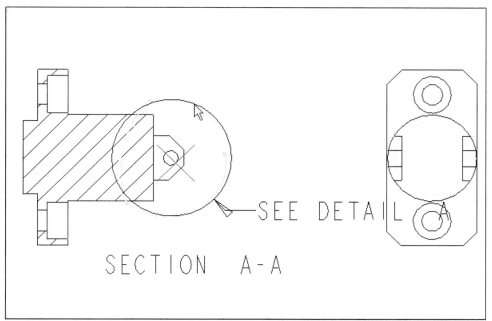

3D CAD A detail view area is defined on the parent view. The identification note and circular border for the detail view are generated by the software.

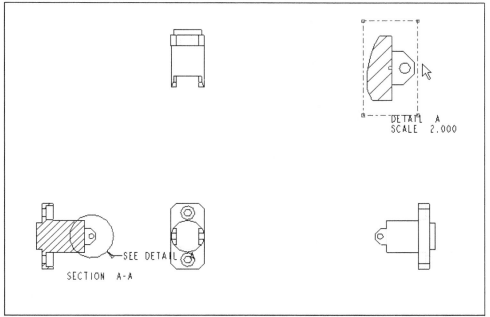

3D CAD Detail views are generated from the parent view, not constructed. The new detail view can be placed and moved about the drawing as desired.

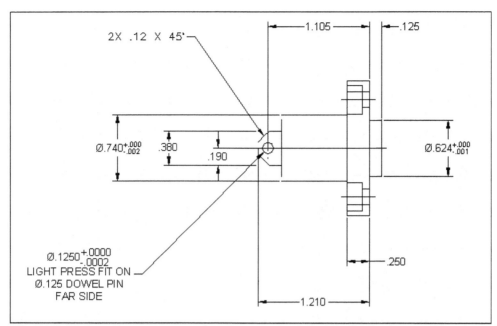

2X .12 X 45'

1.105

.125

Ø.740 +.000 -.002

.380

.190

Ø.624 +.000 -.001

Ø.1250 +.0000 -.0002
LIGHT PRESS FIT ON
Ø.125 DOWEL PIN
FAR SIDE

.250

1.210

2D CAD Hole callouts are completed as required. In 3D CAD, the modification of a hole on the model will automatically update the hole and its note on any drawing where it is displayed.

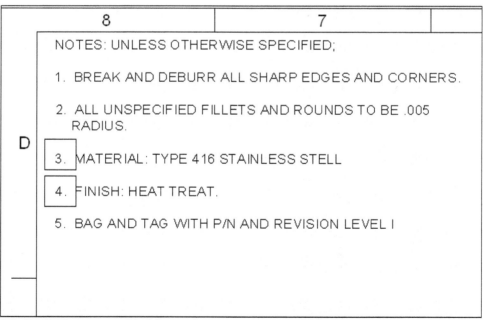

8	7	

NOTES: UNLESS OTHERWISE SPECIFIED;

1. BREAK AND DEBURR ALL SHARP EDGES AND CORNERS.

2. ALL UNSPECIFIED FILLETS AND ROUNDS TO BE .005 RADIUS.

3. MATERIAL: TYPE 416 STAINLESS STELL

4. FINISH: HEAT TREAT.

5. BAG AND TAG WITH P/N AND REVISION LEVEL I

D

2D CAD Notes are added by typing text or as saved blocks

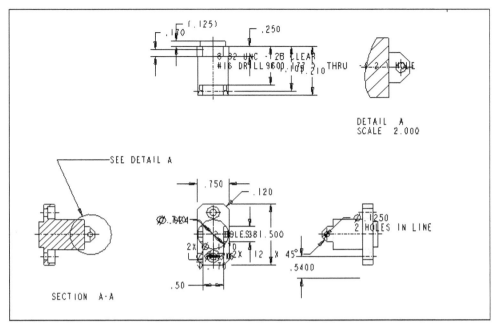

DETAIL A
SCALE 2.000

SEE DETAIL A

SECTION A-A

3D CAD Dimensions and centerlines are displayed with one command

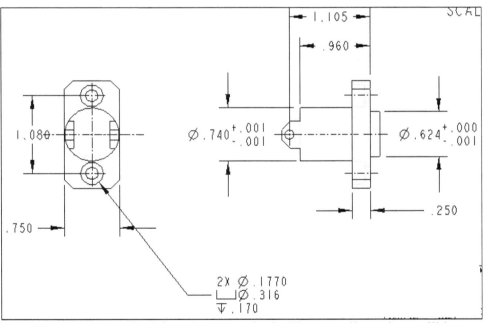

3D CAD Notes are edited for correct standards. The note dimensions will be parametric; therefore, any changes to the hole will be reflected in the drawing hole note. A displayed dimension can be moved to any appropriate view as needed.

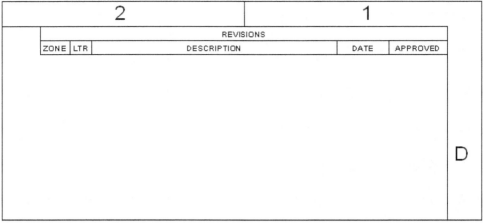

2D CAD Revision block

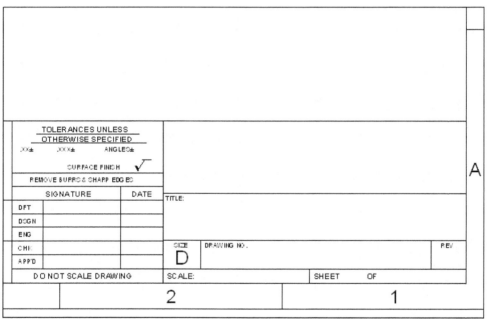

2D CAD Title block accompanies the drawing format similar to all CAD programs. Here the block is still empty.

Annotations

You can annotate the drawing with notes, manipulate the dimensions, and use layers to manage the display of different items on the drawing. **Drawing mode** in parametric 3D design provides you with the basic ability to document solid models in drawings that share a two-way associativity.

```
UNLESS OTHERWISE SPECIFIED DIMENSIONS ARE IN INCHES
DECIMALS .XX±.01  XXX±.005      ANGLES±.5°

REQUIREMENTS FOR FINISHED PARTS UNLESS OTHERWISE SPECIFIED
ALL DIMENSIONS APPLY AFTER COATING
ON ALL PLATED PARTS CORNERS MUST BEND SMOOTHLY

TOLERANCES FRACTIONAL.............. ±1.64
CORNER BREAK (EXTERNAL)(DEBUR)..... .002 TO 0.20R (OR EQUIVALENT)
FILLET RADII (INTERNAL)............ .002 TO 0.20R
THREAD CHAMFER (OR CSK)............ 1 TO 1/2 THREADS DEEP
SURFACE INTERSECTIONS.............. ±.010 MAX
HOLES AND FEATURES TO BE ON THE ℄. ±.010 MAX
GENERAL FINISH..................... 63/ MAX

◎   CONCENTRICITY      .006 TIR
    PERPENDICULARITY   .006 IN/IN TOTAL (MAX)
    PARALLELISM        .010 (MAX)
    ROUNDNESS          .006 (MAX)
    FLATNESS           .006 IN/IN (MAX)
    STRAIGHTNESS       .001 IN/IN (MAX)
    SYMMETRY           .020 (MAX)
                                          THIRD ANGLE
                                          PROJECTION
```

3D CAD Common notes and specifications are imbedded on the format, not the drawing. Specific notes and tolerances are added to the 3D model. Your customized format will have parametric fields that will automatically read the models name, notes, tolerances, scale, drawing name, sheet number, etc. when it is added to an existing drawing sheet.

```
THIS DOCUMENT CONTAINS CONFIDENTIAL AND PROPRIETARY INFORMATION OF CAD RESOURCES
NEITHER THIS DOCUMENT NOR THE INFORMATION HEREIN MAY BE REPRODUCED, USED, OR DISCLOSED TO OR
FOR THE BENEFIT OF ANY THIRD PARTY WITHOUT THE PRIOR WRITTEN CONSENT OF STRYKER.

TITLE:

          PEDESTAL DESIGN

SIZE    DWG. NO.                                        REV.
  B                PEDESTAL

    DO NOT SCALE DRAWING          SHEET 1 OF 1
```

3D CAD Title block information from the model and the drawing is "read" into your customized format as it is applied to the drawing sheet. Drawing templates can be created to streamline this process.

NOTES: UNLESS OTHERWISE SPECIFIED:

1. BREAK AND DEBURR ALL SHARP EDGES AND CORNERS

2. ALL UNSPECIFIED FILLETS AND ROUNDS TO BE .005
   RADIUS

3. MATERIAL: TYPE 416 STAINLESS STELL

4. FINISH: HEAT TREAT

5. BAG AND TAG WITH PIN AND REVISION LEVEL I

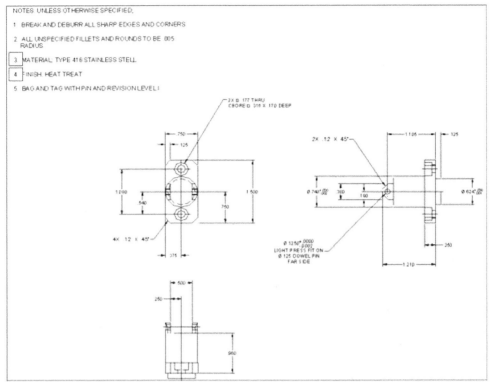

2D CAD Views and dimensions are completed

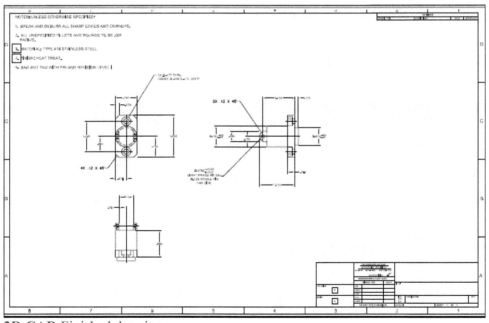

2D CAD Finished drawing

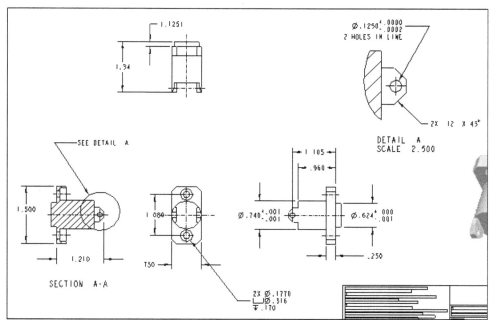

3D CAD Detailing is complete

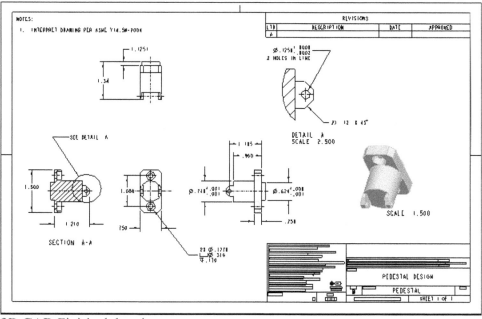

3D CAD Finished drawing

51

3D CAD and design changes in the drawing

Changes that are made to the 3D model in Part mode or Assembly mode will cause the drawing to update automatically and reflect the changes. *Any changes made to the 3D model in the Drawing mode will also be immediately visible on the 3D model in the Part and Assembly modes.* This is one of the biggest benefits of 3D CAD. In 2D CAD, make a change to the design and you have to update individual entities and views manually.

Using 2D CAD, it's nearly impossible to develop various configurations of products, assemblies, or families of products efficiently, since each individual part, assembly, and drawing must be redrawn from scratch. Some 3D CAD systems offer configuration management tools, which enable users to create multiple variations of a product in a single document. These tools also help users to develop and manage families of parts and models with different dimensions, components, properties, and other parameters.

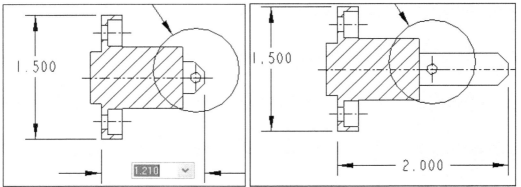

3D CAD Modifying the 1.210 dimension to 2.000 will cause the part model to update after regeneration

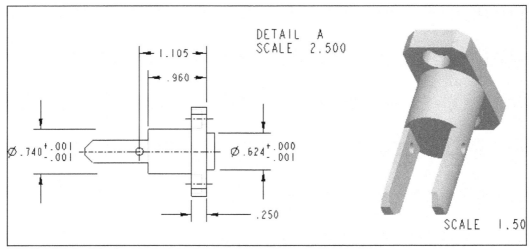

3D CAD Modified component

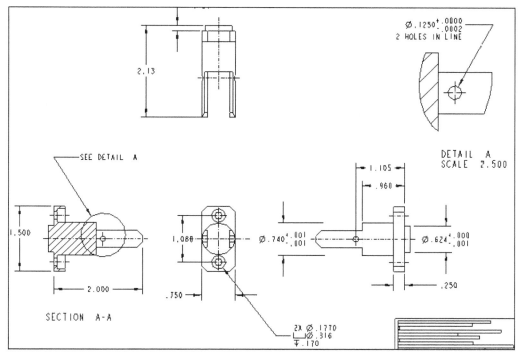

3D CAD The model updates all the views of the drawing

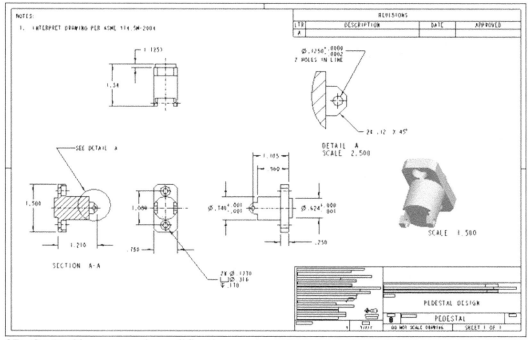

3D CAD Changes and modifications are undone and the original part and drawing specifications are displayed

3D CAD and manufacturing

The database created when modeling the part is used directly in the manufacturing module. In many 3D CAD systems, the database must be ported to outside manufacturing programs before the machining sequence can be created. This porting of internal CAD data to an external CAM program can cause a number of problems and of course you must purchase a different program at great expense. (Of course models created for an integrated part database can also be used in external CAM programs if your facility is geared to manufacture with an existing CAM system.)

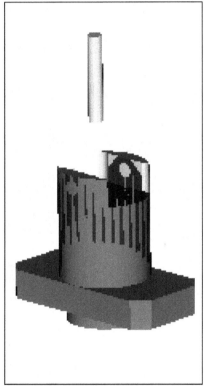

3D CAD Milling

In the accompanying tutorial there are (a set of) steps provided to allow you to experience this seamless process.

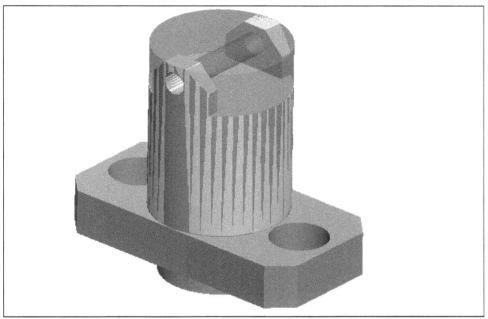

3D CAD Manufacturing model

3D CAD The tool and tool path can be displayed when running the machining simulation

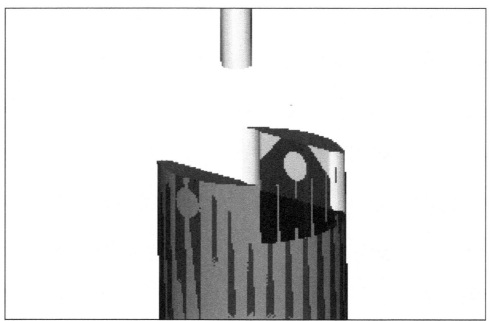

3D CAD Material removal can be displayed when running the machining simulation

If an ECO has indicated a change in the model, these changes will affect the component and manufacturing model. All 3D CAD features are parametric and you have a unified database therefore the manufacturing model will also be modified and the milling sequence will update as well. This is another extremely important factor when choosing a 3D CAD system. If the program uses an outside CAM program, any changes to the CAD model will not propagate to the manufacturing model and a considerable amount of time and effort will be required to instigate a change at this stage of the project.

Chapter Four Comparing the workflow: Assemblies and documentation

- Top-down and bottom-up design processes
- Assemblies and uses
- Laying out an assembly with 2D CAD
- Assemblies created in 3D CAD

Just as you can combine features into parts with a 3D CAD system, you can also combine parts into assemblies. 3D CAD enables you to place component parts and subassemblies together to form assemblies. You can modify, analyze, or reorient the resulting assemblies.

Exploded views, sections, and a variety of display options make the assembly model created in 3D CAD usable to generate any number of formally complicated and time-consuming procedures associated with 2D constructions. Technical illustrations, historically bordering on "art" because of the required techniques, etc. can be derived directly from the 3D assembly model, therefore saving time and increasing their usefulness. Using the Internet/Intranet, 3D designs can be easily shared throughout your company regardless of where the design and the manufacturing take place. Online collaboration can also be utilized so that different aspects of the assembly and its related components are worked on simultaneously. Revisions and ECO's on components can be more easily propagated throughout the assembly in comparison to a 2D rendition of the product which would require drawing changes and assembly drawing changes and the distinct possibility of cascading design inconsistencies.

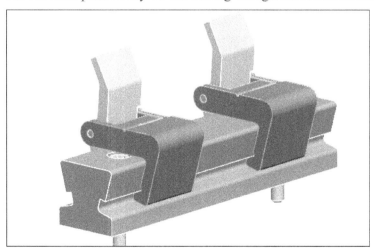

Finger clamp assembly modeled with a 3D CAD system

Note: An online tutorial is available at www.cad-resources.com that will guide you through the steps required to create the assembly. For online files, click www.cad-resources.com > **PTC Partnership** > **Chapter 4 ...** > download and open the zipped support files.

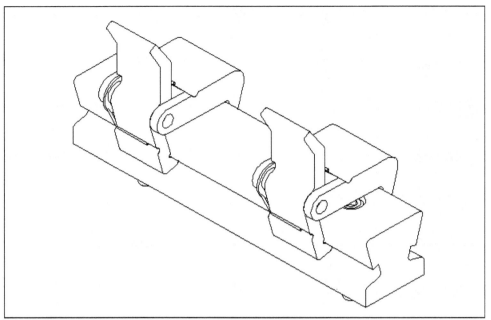

Finger clamp assembly drawn with a 2D CAD system

With 2D assembly drawings, the fit and finish of the components and the assembly are left to the visual 2D paper description. Interfaces cannot be checked, the assembly cannot be used for collision detection, including mechanism and motion analysis, or clearance and interference checking. Design problems cannot be determined at an early stage of development and sometimes not until manufacturing discovers them and their effects on the assembly.

Using a high-level 3D CAD system, you can also create process plans and serviceability documentation for your assemblies such as:

- Create individual steps for the assembly process.
- Add instructions for each step of the process.
- Assemble and Disassemble components.
- Create Fabrication Units.
- Assemble fixtures and tooling to the assembly without altering the Design Assembly.
- Regroup components independently of the design assembly, to model the fabrication structure accurately.
- Create time and cost estimates for each process step.
- Customize the display of each process step by defining multiple explode states.
- Create explode-offset lines and assign different colors and line fonts to components based on their status in a step.
- Create detailed drawings of each step in the process.
- Create a manufacturing BOM for each step in the process.

Another area of increased productivity is associated with assembly drawings and a bill of materials. 3D CAD systems can also create a Bill-of-Materials report (a BOM, or parts list) for an assembly. All the defining parameters and quantities for each component can be automatically compiled into a table, which may be used on an assembly drawing or formatted for use in your company's production control system.

Within a drawing format, a table can be set up to automatically read parameters and quantities of each component in the assembly. When this drawing format is added to an assembly drawing, this table will show a bill of materials (parts list).

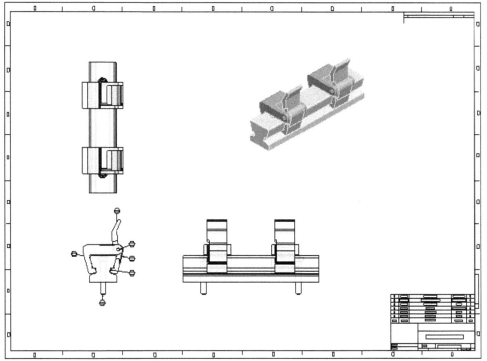

3D CAD Assembly drawing

6	FC-TS-101	TORSION SPRING	SPRING STEEL	2
5	FC-SHCS-101	SOC HD CAP SCREW	STANDARD PART	2
4	FC-RE-101	RAIL EXTENSION	STEEL	1
3	FC-DP-101	DOWEL PIN	STANDARD PART	2
2	FC-CF-101	CLAMP-FINGER	STEEL	2
1	FC-CB-101	CLAMP-BODY	STEEL	2
ITEM	PT NUM	DESCRIPTION	MATERIAL	QTY

3D CAD Assembly drawing parts list (BOM)

By defining parameters in the parts and subassemblies in a 3D CAD assembly, that agree with the specific format of the parts list (BOM), you make it possible for the system to automatically add pertinent data to the assembly drawing as components are added to the assembly.

With the 2D CAD drawing, you are limited to a drawing and a parts list that requires the BOM items be typed manually into a tabular form. Instead of the components' parameters, such as, material, part name, and part number being automatically derived from the component model parameters, the designer must painstakingly type each entry.

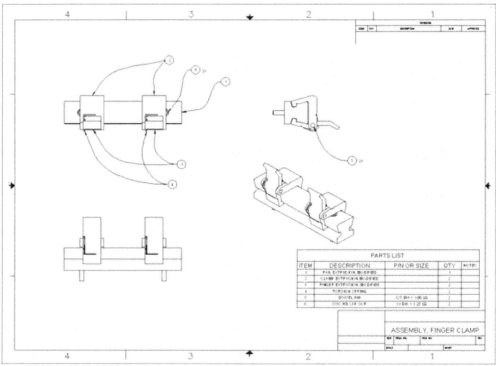

2D CAD Assembly drawing

PARTS LIST				
iTEM	DESCRIPTION	P/N OR SIZE	QTY	NOTES
1	RAIL EXTRUSION, MODIFIED		1	
2	CLAMP EXTRUSION MODIFIED		2	
3	FINGER EXTRUSION, MODIFIED		2	
4	TORSION SPRING		2	
5	DOWEL PIN	.125 DIA X 1.00 LG	2	
6	SOC HD CAP SCR	.19 DIA X 1.25 LG	2	

2D CAD Assembly drawing parts list (BOM)

Assembly design process in 2D and 3D

There are two basic ways to design the components of an assembly: by modeling (3D CAD) or laying out (2D CAD) the parts while working with the assembly, or by modeling (3D CAD) or drawing/detailing (2D CAD) each component separately and then assembling them together.

The first approach is commonly known as top-down design, the second as bottom-up design. These terms describe design processes. In reality most design projects include both top-down and bottom-up procedures. For standard components added to an assembly, regardless in 2D or 3D, are always bottom-up since the items already exist and do not require detail drawings.

2D CAD Top-Down Design

In the **top-down** model an overview of the system is formulated, without going into detail for any part of it. Each part of the system is then refined by designing it in more detail. Each new part may then be refined again, defining it in yet more detail until the entire specification is detailed enough to validate the model. The steps for parts using 2D CAD assembly design top-down procedure:

- Layout the assembly with appropriate views
- Extract the part geometry from the assembly
- Create the views for the part description
- Add the dimensions and appropriate notes
- Add the format and complete the title block

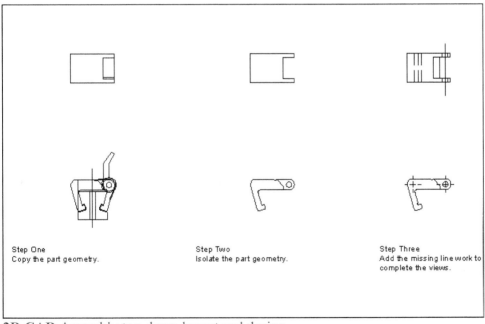

Step One
Copy the part geometry.

Step Two
Isolate the part geometry.

Step Three
Add the missing line work to complete the views.

2D CAD Assembly top-down layout and design

3D CAD Top-Down and Bottom-Up Design

Top-down design allows you to create parts in the context of the product assembly, referring to existing geometry as you create new part features. Bottom-up design is used to create an assembly that is comprised mostly of purchased parts, and parts designed for use in earlier products, or whenever you are reverse-engineering a product.

Bottom-up design—each designer creates a set of detail parts, each of which has its own stand-alone part file. The assembly file is created after the detail parts are finished, by assembling the detail parts and (purchased) standard components (parts) together. In bottom-up design, first the individual parts of the system are specified in great detail. The parts are then linked together to form larger components, which are in turn linked until a complete system is formed. This strategy often resembles a "seed" model, whereby the beginnings are small, but eventually grow in complexity and completeness.

Top-down design—each designer creates part and assembly files, using or referencing files from other designers as required. This situation often requires that two or more people work on an assembly simultaneously; so good data management practices are required. Top-down design is used to design detail parts within the context of an assembly, which is typical of most design situations. It is very easy to build parametric links between assembly components with this approach; for example, aligning a sketched entity for a cut in one component to an edge in another component. The relationship created is referred to as an external reference. External references should be used with care, since missing external references are a common source of regeneration failures. The following describes one version of steps for top-down design:

- Create an assembly file.
- Create and assemble a company-defined start part.
- Create or import section or part geometry.
- Create and assemble standard and existing components where possible.
- Model the required features in new components using the existing component geometry to drive the varying parameters.
- Activate a drawing and format and establish views.
- Add a format with a parametric title block and BOM table.

3D CAD part mode vs. assembly mode modeling

There are two different methods of working with 3D CAD part and assembly files. To make changes to a design, you may modify the part file by itself in part mode, or modify the part file from within the context of the assembly, in assembly mode. In part mode you are working only with geometry in the part, and are working in a window that contains only the part. In assembly mode, you are manipulating the assembly, and can be working with geometry in the assembly or in one of the parts.

With 3D CAD when a component used in an assembly changes, i.e., a dimension is modified or a feature added, the changes will be propagated (visually) in the assembly. This is true when the part is opened by itself and changed, or when it is changed within the context of the assembly.

This is yet another example of associativity (bi-directional flow of information). It is important to realize that there is only one model of a part. That model is then referenced/linked (not copied) wherever it is used in the design, documentation, and manufacturing process. Only systems that have built-in CAM modules, allowing the part database to be used directly in the manufacturing, NC, etc. mode will have this dynamic bi-directional associativity.

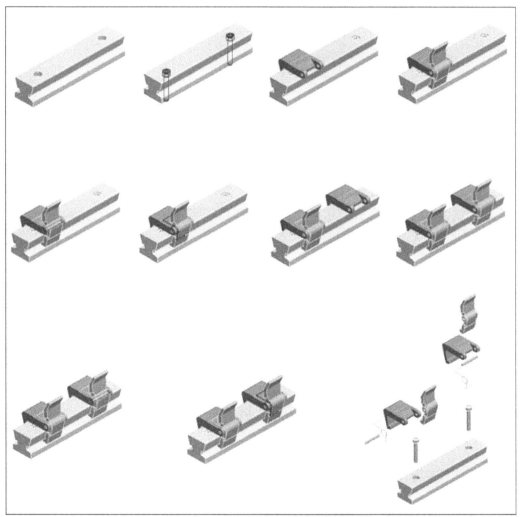

3D CAD Assembly

Assembly design process in 2D (left pages)

Top-down design will incorporate files that contain a wide variety of design information including, but not limited to: tolerances, interfaces, relations, parameters, tables, graphs, and dimensions.

In our simple example, the design layout incorporates the rail section and its relationship to the components in the assembly. The rail cross-section can be imported or created in a 2D CAD drawing/sketch or in 3D CAD as a section. The geometry can be published and thereby available to use as controlling geometry across a variety of parts and assemblies as needed.

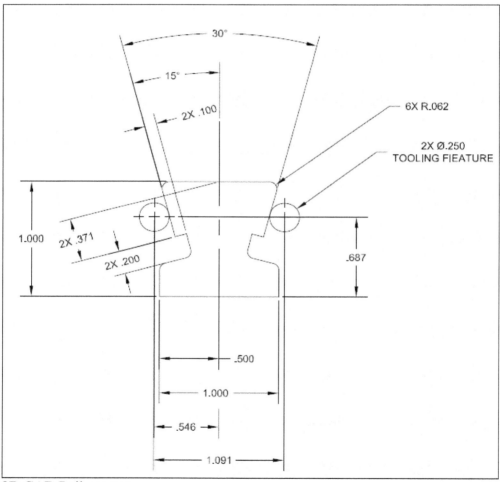

2D CAD Rail geometry

Assembly design process in 3D (right pages)

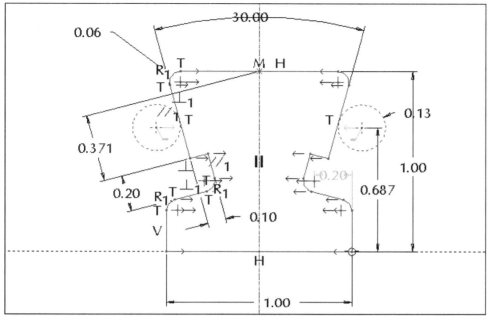

3D CAD Rail section geometry

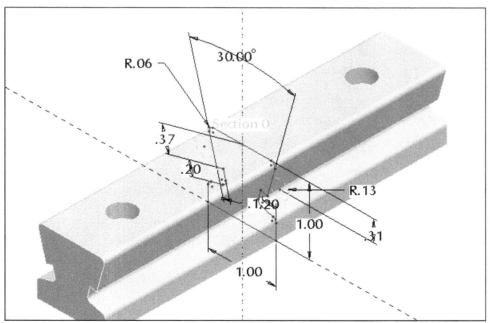

3D CAD Section geometry used to create the model of the rail extrusion

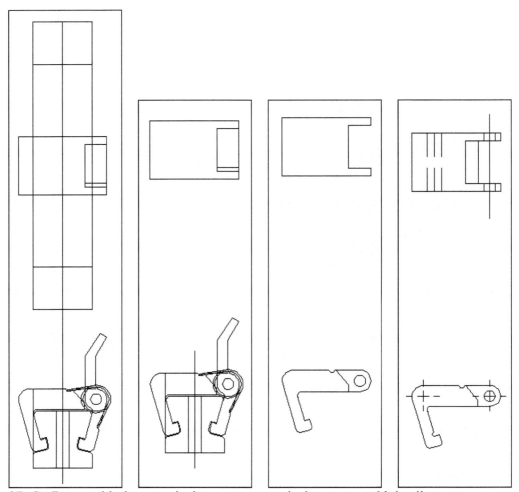

2D CAD assembly layout – isolate geometry – isolate part – add details

The 2D CAD design sequence involves laying out two views of the assembly, isolating individual components from the assembly geometry, and detailing the parts. The original layout serves as the controlling geometry for part geometry where it relates to the rail section.

The 3D CAD design sequence starts with the primary driving rail part. The rail sectional geometry is created or imported and the component is modeled as per the design. The second component, the clamp body, could be modeled as a unique part and added to the assembly, or created in a top-down design methodology that references the rail geometry. This second method creates external references within the clamp body (references which can later be removed and rerouted).

The top-down method can therefore be implemented to control the geometry of related parts and used to insure proper alignments and connections.

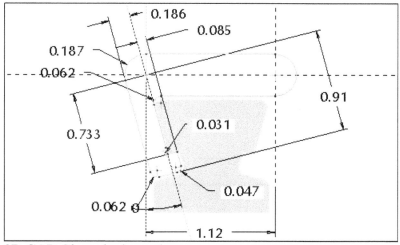

3D CAD Clamp body section geometry

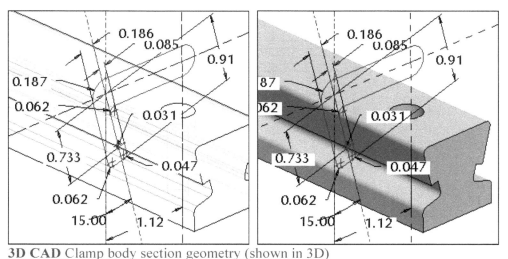

3D CAD Clamp body section geometry (shown in 3D)

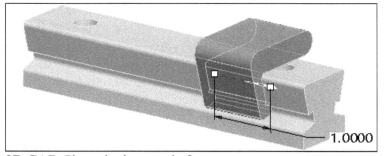

3D CAD Clamp body extrude feature

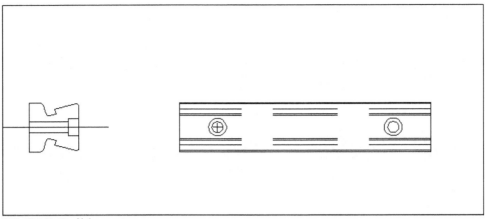

2D CAD Rail layout

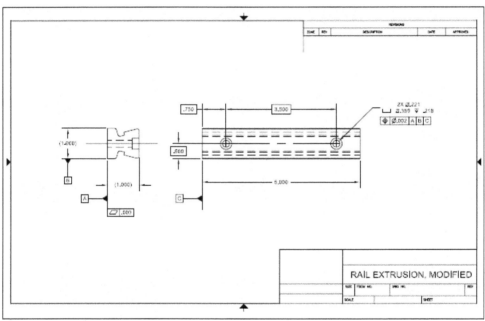

RAIL EXTRUSION, MODIFIED

2D CAD Rail detail

The 2D CAD details are then initiated starting with the rail. For the 3D CAD method, each of the components making up the design will need to be modeled individually, modeled using top-down methodology, or imported from a standard parts library/online catalog (the screw and dowel).

The components are then sequentially added to the assembly, one by one. In the example, the rail and screws are assembled. In normal practice, the clamp body, finger, spring, and dowel would be established as a subassembly before adding the full clamp to the primary assembly.

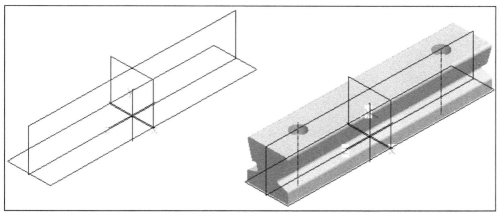

3D CAD Assemble the rail

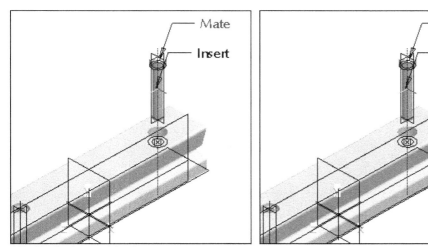

3D CAD Assemble the screws

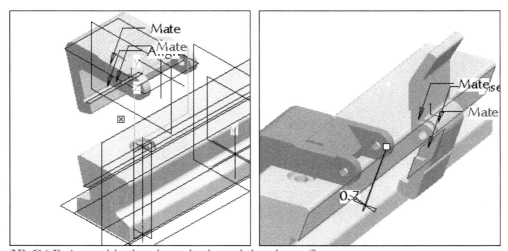

3D CAD Assemble the clamp body and the clamp finger

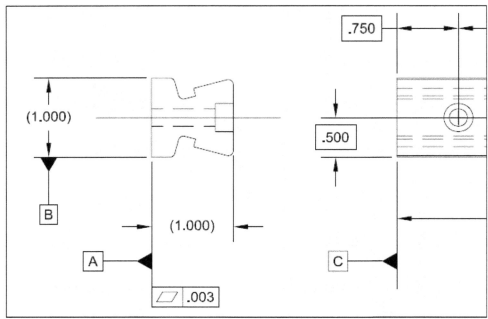

2D CAD Rail dimensions side view

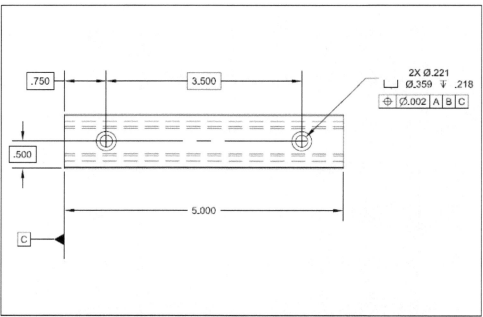

2D CAD Rail dimensions front view

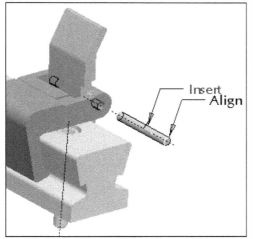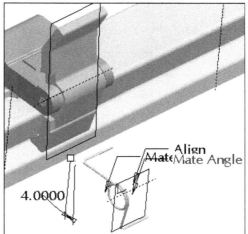

3D CAD Assemble the pin and the spring

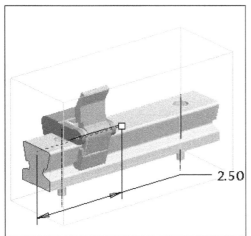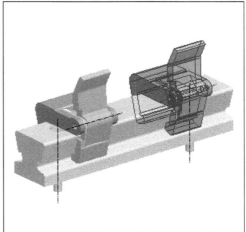

3D CAD Clamp is moved and copied

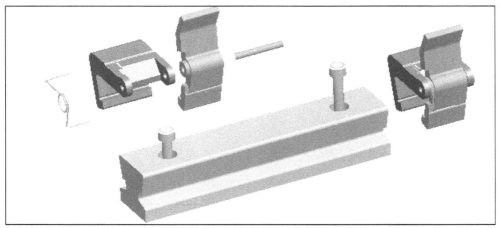

3D CAD Pictorial and exploded view created from the assembly model

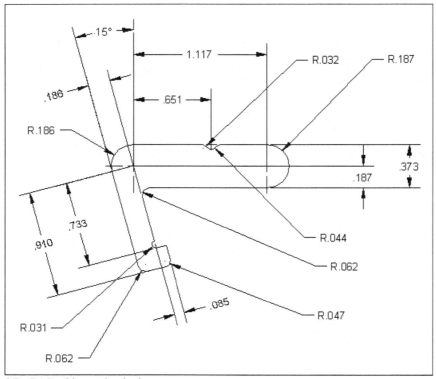

2D CAD Clamp body layout

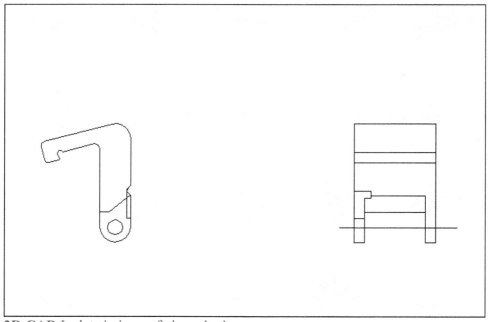

2D CAD Isolated views of clamp body

Each component, including ones taken directly from a standard parts catalog or library catalogs, requires that parameters representing items for the BOM be input and saved with the model. The parameter names correspond to identically named parameters that are used in the BOM table that is created for the drawing format. As the assembly model is added to a drawing and a format added to the drawing, the BOM is automatically completed by reading in the component parameters into the table appearing on the drawing.

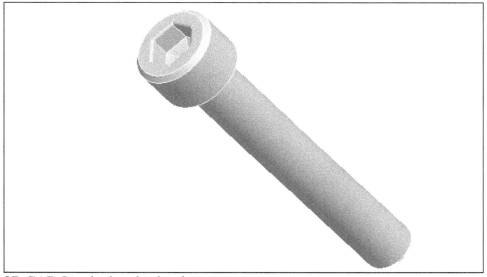

3D CAD Standard socket head cap screw

Name	Type	Value	Designate	Access	Source
BASIC_DIA	String	NO.10	☐	🔒Full	... User-Def.
MAT	String	STANDARD PART	☐	🔒Full	... User-Def.
DSC	String	SOC HD CAP SCREW	☐	🔒Full	... User-Def.
PRTNO	String	FC-SHCS-101	☐	🔒Full	... User-Def.

3D CAD Part parameters are imbedded in the component

The socket head cap screw may have some parameters already imbedded by the manufacturer. Here, the Name: **BASIC_DIA** and Value: **NO.10**. The MAT, DSC, and PRTNO are input by the designer. These parameters will later be displayed as per their related assignments on the drawing format BOM table.

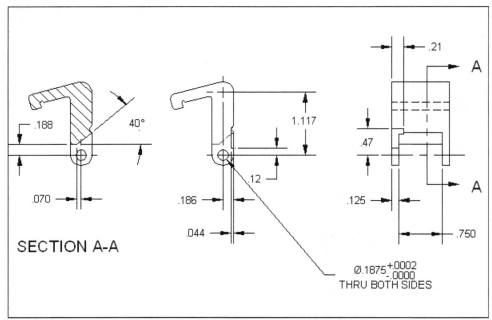

2D CAD Detailed views of clamp body

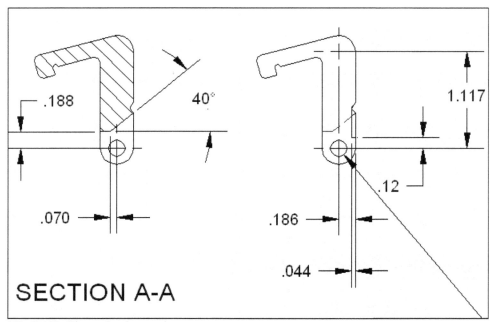

2D CAD Clamp body section

74

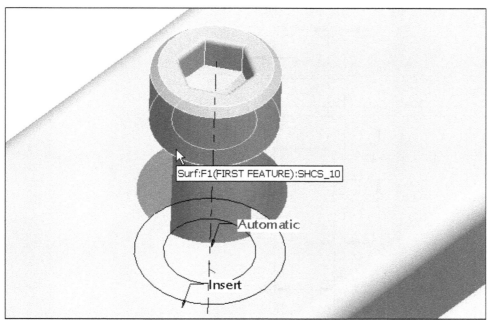

3D CAD Screw being assembled

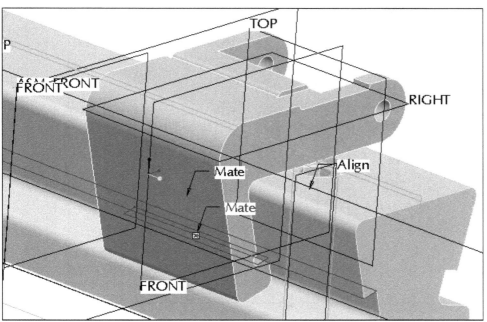

3D CAD Clamp body being assembled

As a component is added to the project, assembly constraints are utilized to position and lock the component in place.

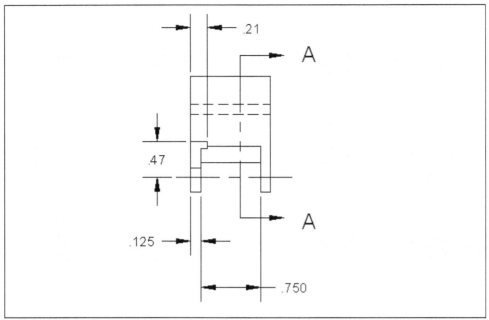

2D CAD Clamp body view

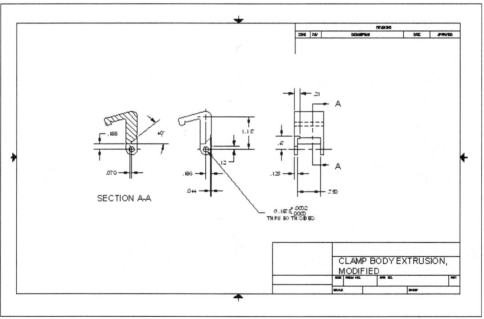

2D CAD Clamp body detail

&rpt.index	&asm.mbr.PRTNO	&asm.mbr.DSC	&asm.mbr.MAT	&rpt.qty
ITEM	PT NUM	DESCRIPTION	MATERIAL	QTY

TOOL ENGINEERING CO.

DRAWN		DRAWING SCALE	ASM_FORMAT_E	
ISSUED				SHEET 1 OF 1

3D CAD Formatted title block and parts list table

The title block with repeat areas is established for parameters. In this example, the component parameters include: **MAT** (for material), **DSC** (for description), and **PRTNO** (for part number). Every component in the assembly has these same parameters (with different values of course). The BOM table will automatically propagate with every component and their associated parameters.

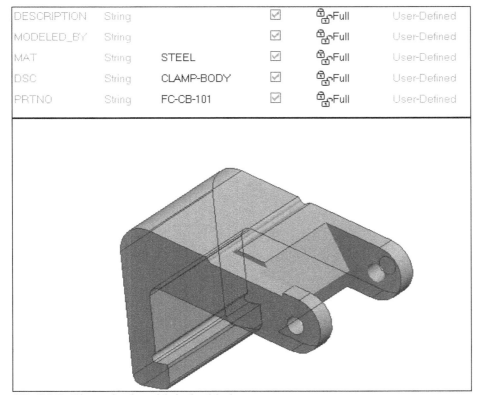

DESCRIPTION	String		☑	🔒Full	User-Defined
MODELED_BY	String		☑	🔒Full	User-Defined
MAT	String	STEEL	☑	🔒Full	User-Defined
DSC	String	CLAMP-BODY	☑	🔒Full	User-Defined
PRTNO	String	FC-CB-101	☑	🔒Full	User-Defined

3D CAD Clamp body with imbedded parameters

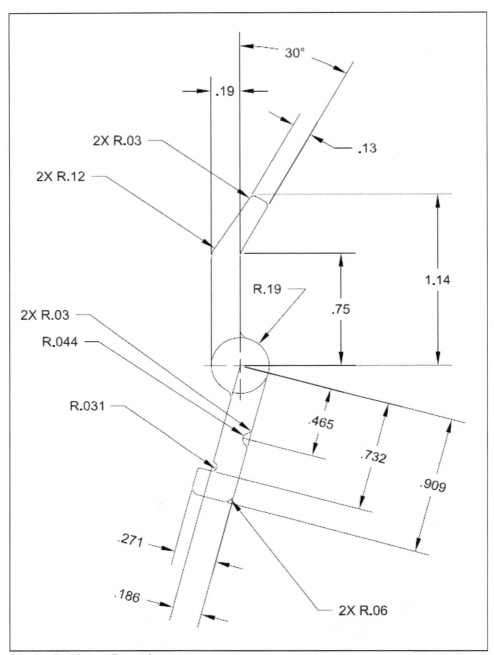

2D CAD Clamp finger layout

For the 2D CAD system, the cross section geometry for the clamp finger is established by the original assembly layout.

For the 3D CAD design, the component can be modeled independently or in the context of the existing geometry and parts of the assembly.

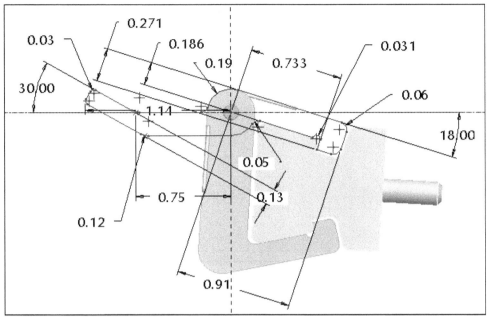

3D CAD Clamp finger section geometry in relation to the assembly parts

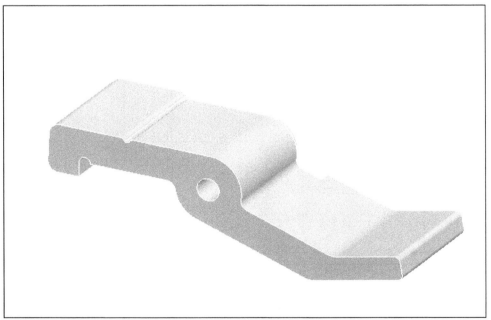

3D CAD Clamp finger

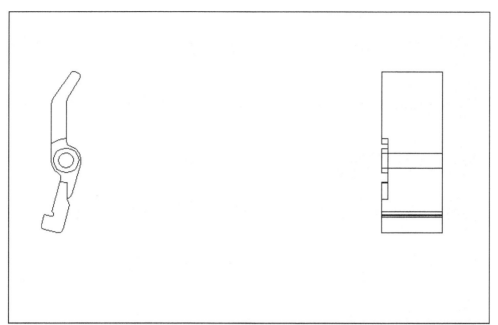

2D CAD Isolated views of the clamp finger

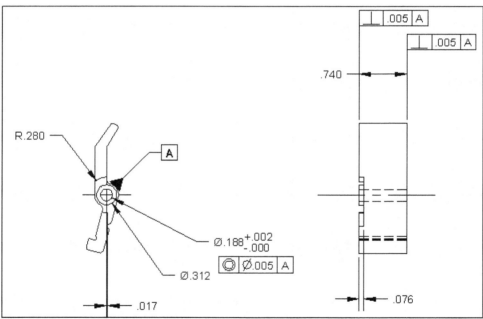

2D CAD Front and side views of the clamp finger

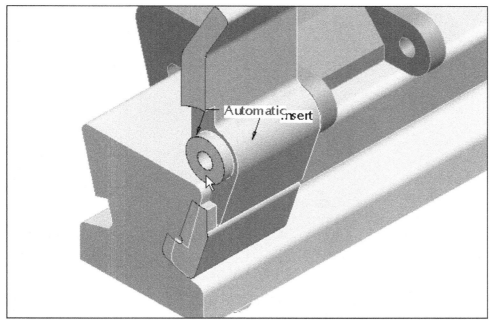

3D CAD Clamp finger being assembled

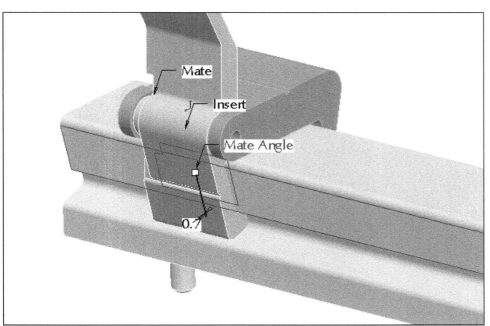

3D CAD Clamp finger assembled to the rail and to the clamp body

81

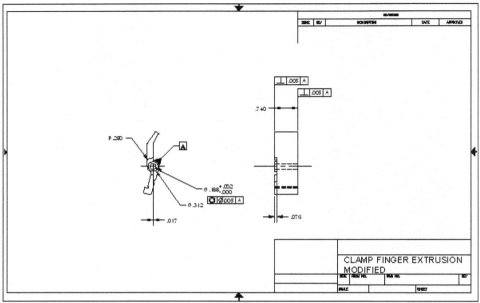

2D CAD Clamp finger detail drawing

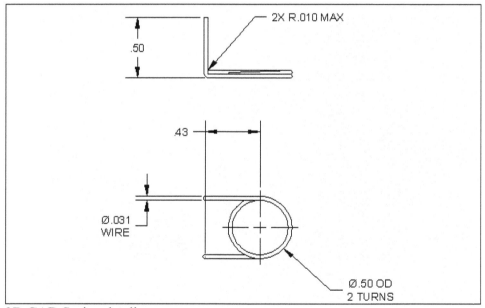

2D CAD Spring detail

The spring is "drawn and detailed" for the 2D CAD system. For the 3D CAD system, the procedure is quite different. When possible, a standard spring is downloaded from an online catalog. If the spring needs to be customized, the model can be redefined and altered. The resulting 3D CAD spring can then be submitted online back to the manufacturer to create the required design.

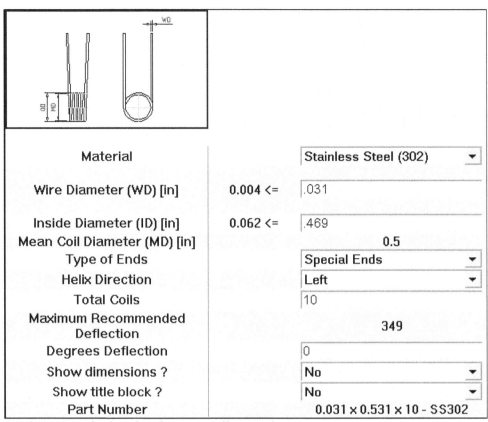

Material		Stainless Steel (302) ▼
Wire Diameter (WD) [in]	0.004 <=	.031
Inside Diameter (ID) [in]	0.062 <=	.469
Mean Coil Diameter (MD) [in]		0.5
Type of Ends		Special Ends ▼
Helix Direction		Left ▼
Total Coils		10
Maximum Recommended Deflection		349
Degrees Deflection		0
Show dimensions ?		No ▼
Show title block ?		No ▼
Part Number		0.031 x 0.531 x 10 - SS302

3D CAD Standard spring from an online catalog

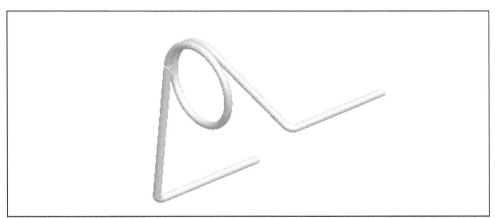

3D CAD Customized spring

When system administrators save standard assembly formats, they will usually include a drawing table that represents the standard parts list. You must be aware of the names of the parameters under which this data is stored so that you can add it properly to your parts.

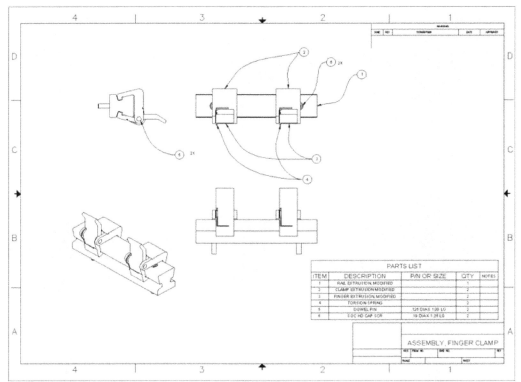

2D CAD Finger clamp assembly

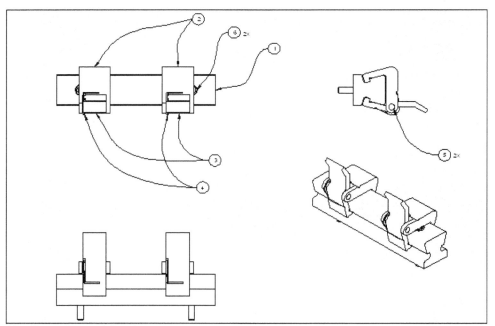

2D CAD Assembly views

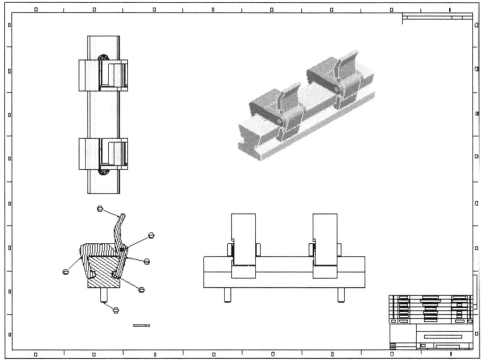

3D CAD Assembly drawing with BOM (front view was changed to display a section in this example)

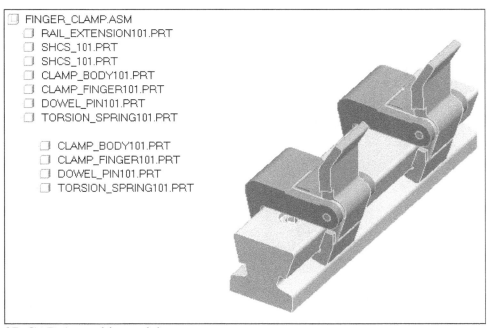

FINGER_CLAMP.ASM
    RAIL_EXTENSION101.PRT
    SHCS_101.PRT
    SHCS_101.PRT
    CLAMP_BODY101.PRT
    CLAMP_FINGER101.PRT
    DOWEL_PIN101.PRT
    TORSION_SPRING101.PRT

        CLAMP_BODY101.PRT
        CLAMP_FINGER101.PRT
        DOWEL_PIN101.PRT
        TORSION_SPRING101.PRT

3D CAD Assembly model

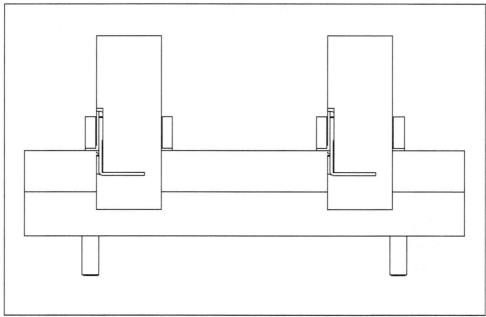

2D CAD Front view

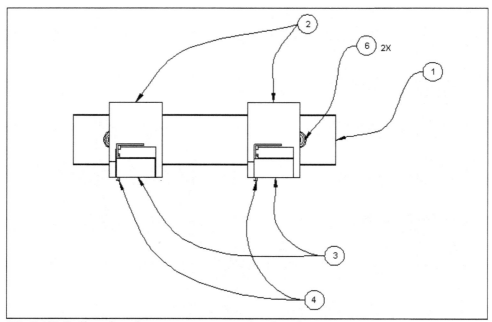

2D CAD Ballooning the components on the drawing

Views need to be constructed individually with 2D CAD. Balloons identifying the assembly components are drawn and positioned manually instead of automatically generated from the component parameters.

As you add components, the system automatically reads the parameters that were established in every component and generates or updates the parts list (BOM). You can also see the same effect by adding these parameters to the components after the drawing has been created.

Part			✓	↖	☐ CLAMP_BODY101			✓

Filter By	All			✓	Sub Items			
Name	Type	Value		Designate	Access	Source	Descri
DESCRIPTION	String			☑	🔒Full	User-Def...	
MODELED_BY	String			☑	🔒Full	User-Def...	
MAT	String	STEEL		☑	🔒Full	User-Def...	
DSC	String	CLAMP-BODY		☑	🔒Full	User-Def...	
PRTNO	String	FC-CB-101		☑	🔒Full	User-Def...	

3D CAD Clamp body parameters

The system also creates BOM Balloons on the first view that was placed on the drawing. You can move these balloons to other views and alter the location where they attach (to the respective parts) to improve their appearance.

6	FC-TS-101	TORSION SPRING	SPRING STEEL	2
5	FC-SHCS-101	SOC HD CAP SCREW	STANDARD PART	2
4	FC-RE-101	RAIL EXTENSION	STEEL	1
3	FC-DP-101	DOWEL PIN	STANDARD PART	2
2	FC-CF-101	CLAMP-FINGER	STEEL	2
1	FC-CB-101	CLAMP-BODY	STEEL	2
ITEM	PT NUM	DESCRIPTION	MATERIAL	QTY

TOOL ENGINEERING CO.

DRAWN		2500	FINGER CLAMP	
ISSUED				SHEET 1 OF 1

3D CAD Bill of materials

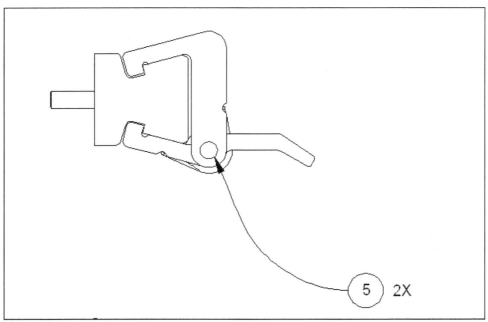

2D CAD Right-side view

PARTS LIST				
iTEM	DESCRIPTION	P/N OR SIZE	QTY	NOTES
1	RAIL EXTRUSION, MODIFIED		1	
2	CLAMP EXTRUSION MODIFIED		2	
3	FINGER EXTRUSION, MODIFIED		2	
4	TORSION SPRING		2	
5	DOWEL PIN	.125 DIA X 1.00 LG	2	
6	SOC HD CAP SCR	.19 DIA X 1.25 LG	2	

ASSEMBLY, FINGER CLAMP

	SIZE	FSCM NO.		DWG NO.		REV
	SCALE				SHEET	

2 | 1

2D CAD Parts list

The parts list (BOM) created with a 2D CAD system is nothing more than typed entries that will hopefully be input correctly. The column headings are simply words that may describe relevant information about the components.

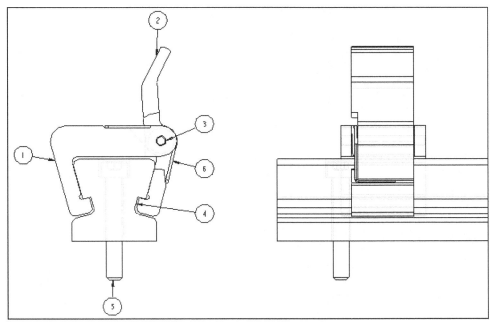

3D CAD Ballooned assembly drawing

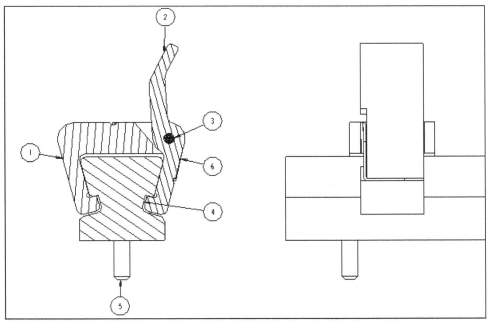

3D CAD Sectioned view and no hidden line display in the front view

Sections can be generated from the existing views depending on the geometry you wish to have emphasized and displayed for clarity.

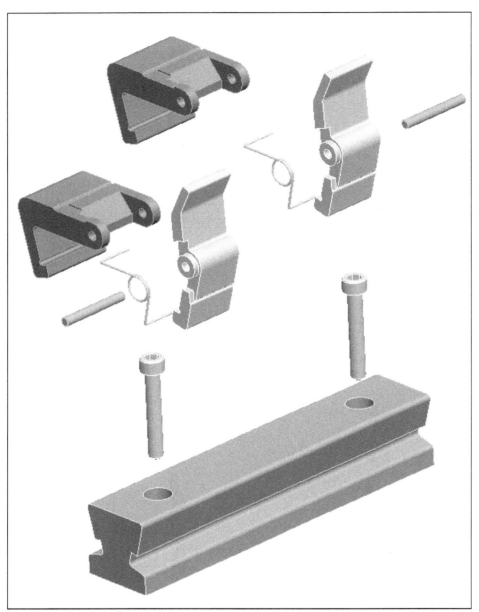

3D CAD Exploded assembly

Exploded views can also be generated from the assembly instead of "drawing" them in isometric or trimetric projection as with 2D CAD. Traditional 2D drawings require complex geometric constructions using difficult procedures to create pictorial projections.

The 3D CAD method can use any number of views created directly from model views and exploded states.

Chapter Five Reusing 2D design data

- How to use 2D data to create a 3D CAD model
- Importing DXF geometry for reuse

One of the most common questions asked by companies contemplating the move from 2D to 3D CAD is "how effectively can I leverage and re-use my existing 2D data? " In this chapter we will explore one option.

With some 3D systems, 2D users are forced to transfer their legacy data to the new system by translating files from one industry-standard format to another. As engineering managers know, this translation introduces substantial data inaccuracy, and often wastes valuable design engineering time. Since this neutral format carries forward a minimum of useful information, the design engineer must spend hours repairing the translated files.

Most of today's modern 3D CAD tools include native translators for AutoCAD DWG/DXF files, and support a range of industry standards including STEP, IGES, CGM, VDA, SET, and VRML, as well as direct translators to all common 3D applications. This helps ensure that the data translation issue is no issue at all.

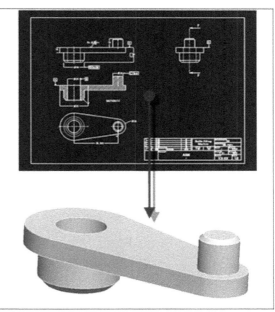

- **Import 2D drawing data**

- **Reference 2D drawing data to create 3D geometry**

- **Benefits:**
  - Easy access to legacy data
  - Reuse legacy data for future designs
  - Leverage 2D legacy data into fully featured parametric 3D models
  - Reduce design time

Reusing 2D drawing files to create 3D geometry

Furthermore, the best 3D CAD solutions provide unique automated data translation aids, which deliver a user-friendly approach to leveraging or reusing 2D legacy data. These tools allow designers to select geometry from their existing 2D drawings, which is then used to automatically create the corresponding 3D features, while, at the same time providing the user with simple training on how to use a 3D system.

Reusing 2D Data

The ability to reuse existing 2D CAD data in future designs makes the transition to 3D CAD easier. Many competent 3D CAD designers would dispute the necessity of conversion, choosing to model the parts and assemblies with native 3D CAD commands. This allows the 3D CAD models to incorporate company standards as well as ensure the utilization of every aspect of the 3D CAD system. But, having multiple methods available to convert your 2D data is an asset.

This chapter covers using 2D CAD data to generate 3D CAD models. In the next chapter, we will use another tool called AutobuildZ (a free PTC add-on). Regardless of which method you and your company choose, knowing that existing designs can be utilized in the 3D CAD program you choose to implement is important.

In general, the greatest benefit will come from using complex shapes that have views usable as 2D sketch geometry when imported as files into sections when creating solid features. Simple geometry will have less value since the 3D CAD modeling process is easily mastered for most basic shapes. On the other hand, when you have a complex geometric shape that was "drawn" in 2D, it may save time importing that geometry and using it to create features in 3D.

We will explore the process of opening an existing 2D CAD drawing into a 3D environment. The drawing is in DXF format. The resulting geometry will be captured view-by-view and used to establish sketch geometry. The sketch geometry will then be used to generate solid features; instead of sketching new geometry as would normally be the practice. In this example, the Arm will be generated from 2D CAD geometry imported into a 3D CAD program.

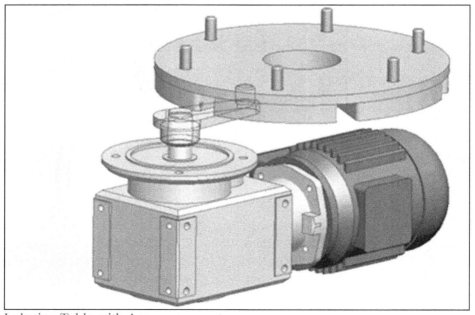

Indexing Table with Arm component

Recreating in 3D using imported 2D geometry

A variety of methods can be used to create 3D CAD geometry, including: remodeling the part from scratch, using imported 2D geometry to drive sketch geometry, or utilizing automated programs such as AutobuildZ which will be overviewed in the next chapter. This chapter will cover using 2D imported data to create 3D component geometry. The accompanying tutorial will guide you through the process, if you choose to experience the process first hand.

The cost of conversion verses remodeling must be kept in mind. For each project, your design team needs to ask the questions: How much of the legacy data needs to be available in 3D?, How long will it take to remodel or convert the data?, What is the cost in terms of project hours to do the conversion?, and finally What are the benefits of having the previous designs available for future projects, ECO's, new product lines, etc.?

The 2D CAD drawing file, Arm_1.DXF will be used to create a fully featured, 3D, parametric part. The existing 2D information can be imported to establish and control the 3D CAD solid features without using new "dimensioned" sketches.

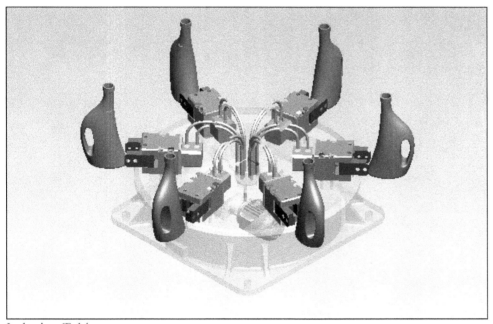

Indexing Table

Note: An online tutorial is available at www.cad-resources.com that will guide you through the steps required to create the Arm part. The tutorial assumes that you have not used Pro/ENGINEER Wildfire 3.0 previously. For online files, click www.cad-resources.com > **PTC Partnership** > **Chapter 5 ...** > download and open the zipped support files.

The arm that connects the motor and the indexing gear needs to be created. This arm has been previously designed using 2D CAD and is available as a drawing file (DXF).

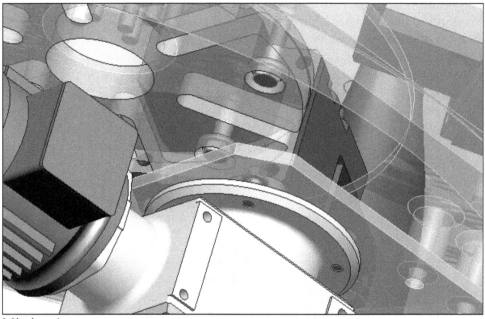
Missing Arm component

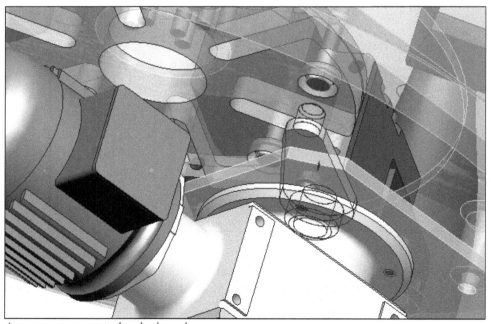
Arm component to be designed

We will use the front and the top (not the hatched cross-section) views drawn using 2D CAD. The front view geometry will be placed on a datum plane and used to create the part's outline (primary feature), the large boss, and the hole. The strategy for modeling the part's features from existing 2D data will be different for every situation. In general, the largest and most important feature will be created first since this will be the "parent" of all succeeding features.

The top view will be placed on a perpendicular datum plane and used to create the smaller boss and chamfer, and as a depth guide for the Arm's thickness, the large boss's height, and the hole's depth.

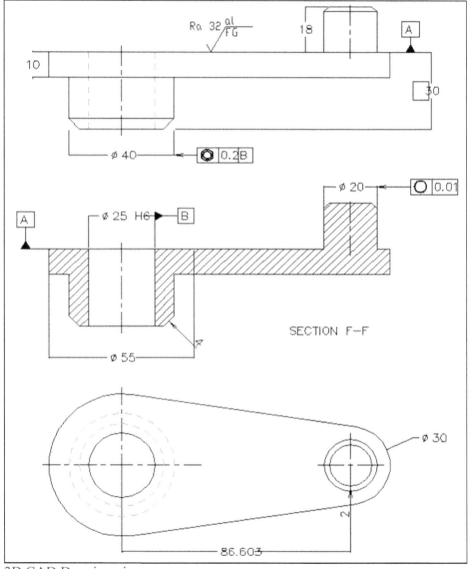

2D CAD Drawing views

The first step is to create a new part using 2D drawing data as the basis for our fully featured 3D parametric part. Then, we import views from 2D legacy data to create geometry in the form of sketches. The process starts with opening the 2D drawing from a file. The unneeded drawing elements are hidden (see the tutorial available at www.cad-resources.com > PTC Partnership > Chapter 5 Tutorial).

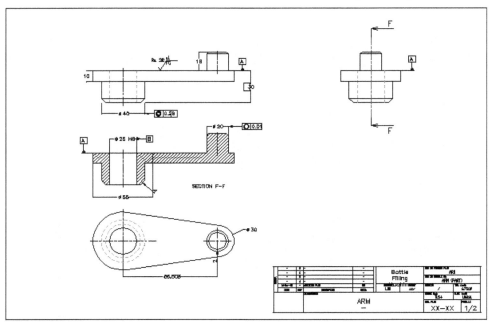

2D CAD Format of imported drawing

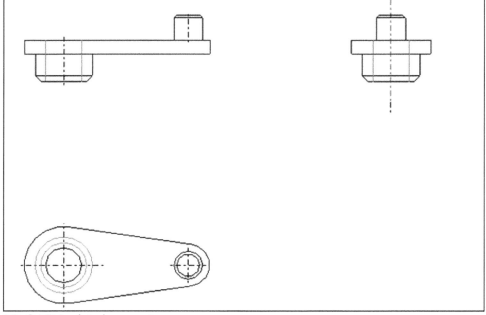

2D CAD Isolated geometry

Next, we create fully featured parametric models from the 2D legacy data. The imported sketch can now be changed and will update intelligently based on the constraints placed by the (3D) Sketcher tool. Continue to import more legacy data as needed to create the 3D part. The 2D legacy data's various cross sectional views can be placed on top of each other, and they can be used as references for exact duplication to the 3D part. A new 3D part is started and the TOP datum plane is selected for the new sketch geometry.

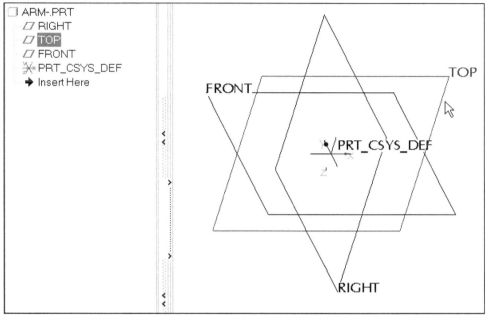

Part modeling started

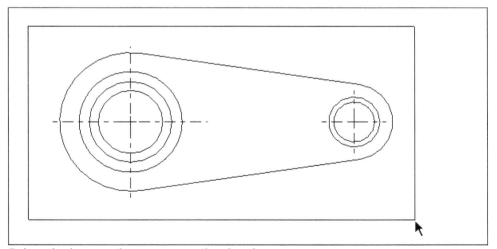

Select the imported geometry as the sketch geometry

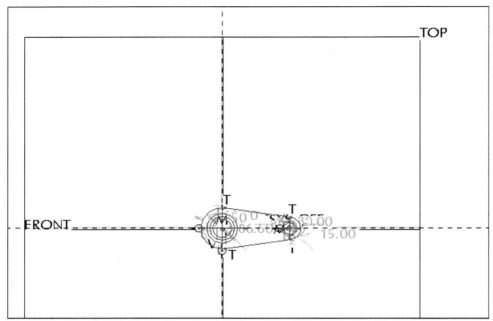

The geometry is transferred to the sketch

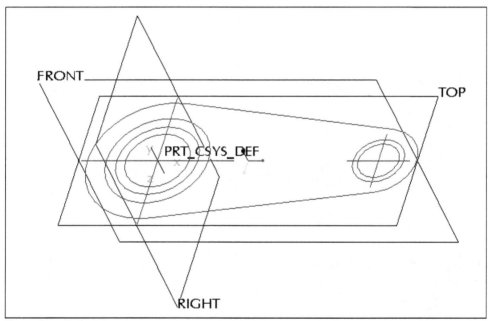

Completed sketch geometry

The second view is now transferred to establish more sketch geometry on the 3D model.

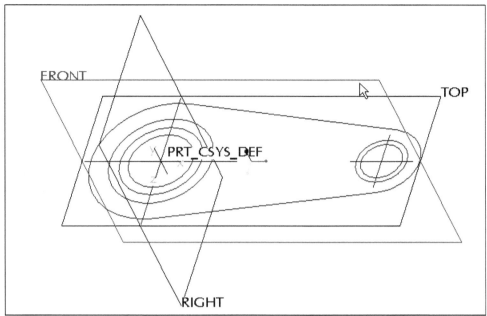

The FRONT datum plane is selected for the next imported geometry

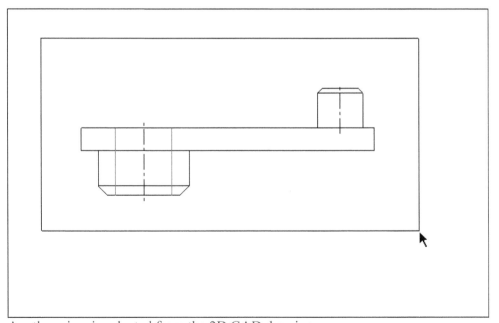

Another view is selected from the 2D CAD drawing

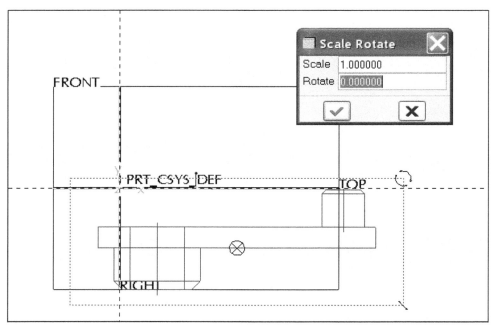

The new geometry is dragged in and dropped at the origin of the part

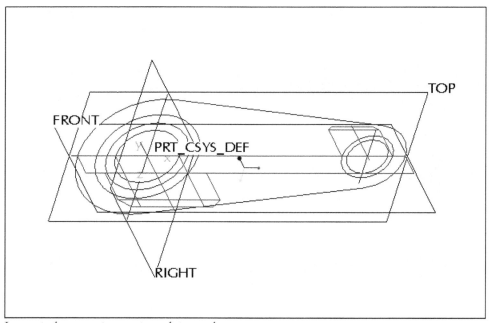

Imported geometry on two datum planes

Remember, you can follow the process in detail using the tutorial or you can do the project yourself with the step-by-step instructions provided in the tutorial available at: www.cad-resources.com > PTC Partnership > Chapter 5 Tutorial. All instructions and supported files are available for download.

The first solid feature will now be created using the imported sketch geometry.

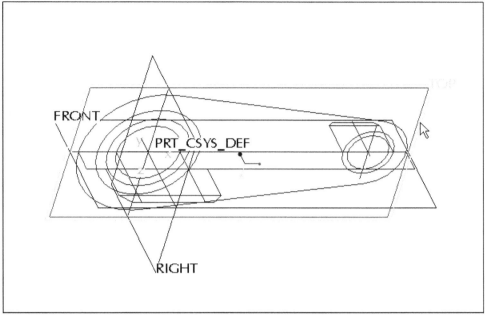

The TOP datum plane is selected as the construction plane

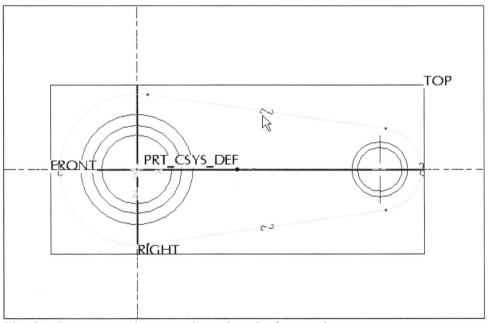

Simple edge commands are used to select the features loop geometry

Sketching first, then extruding the 3D solid from the sketch is a familiar workflow to many 2D users. Here the geometry for the first solid extrusion is created from the imported 2D data.

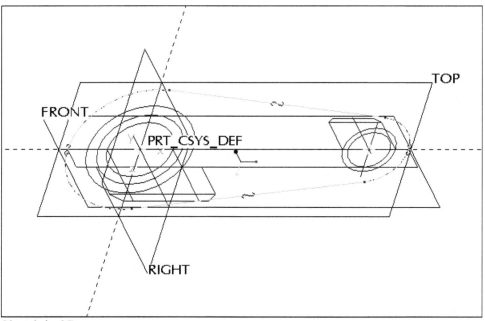

Sketch in 3D

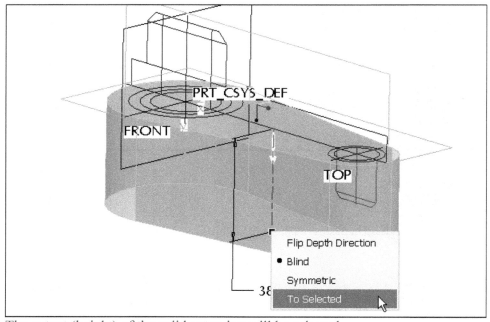

The extent (height) of the solid extrusion will be selected

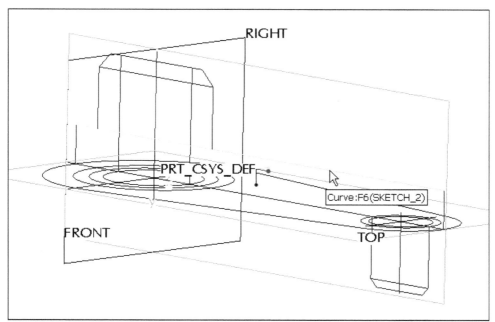

Imported sketch line is selected

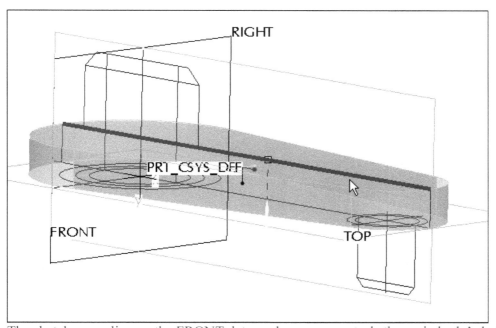

The sketch curve line on the FRONT datum plane now controls the main body's height.

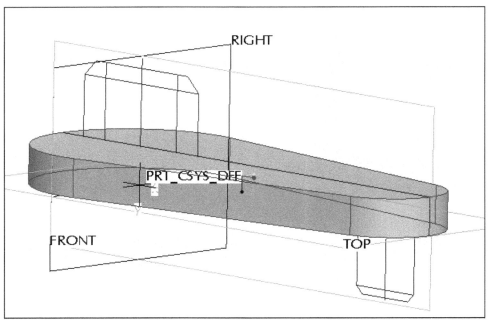

Main body modeled

The next feature is now modeled. The 2D legacy data will again be used as reference for exact duplication.

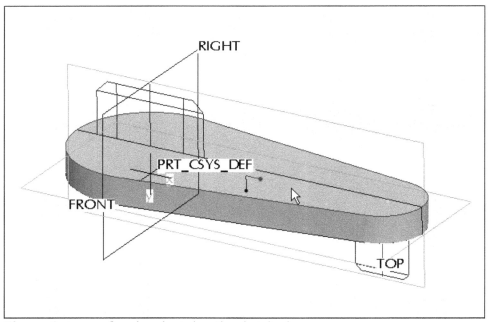

The part's top surface is selected as the sketch plane

The outer circle of the axle will be used to control the new solid feature's diameter.

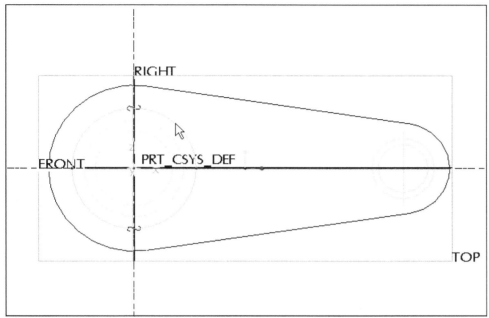

The axle's circle is selected

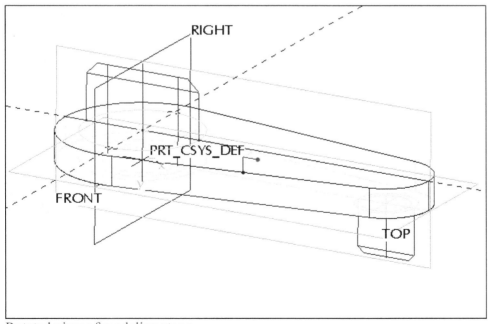

Rotated view of modeling stage

We will use the imported sketch to control the height of the boss.

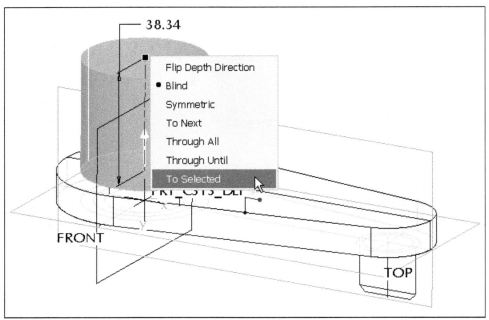

Height of the axel boss will be selected

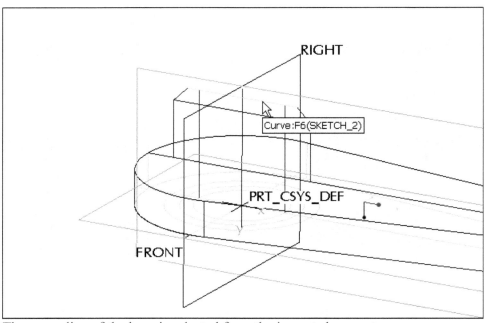

The upper line of the boss is selected from the imported geometry

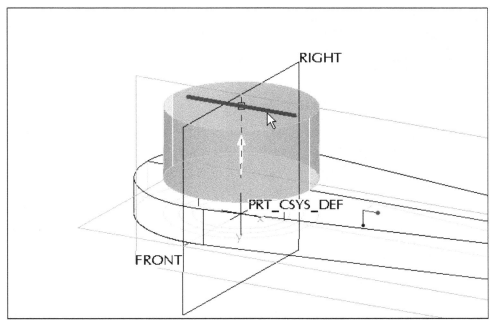

Previewed axel boss

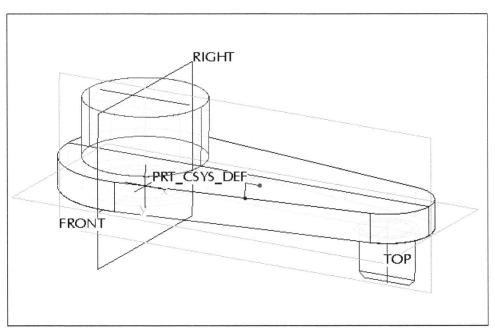

Completed solid feature

A simple pick and place feature will be used to create the chamfer. Pick the edge and then choose the Chamfer command (or vice versa).

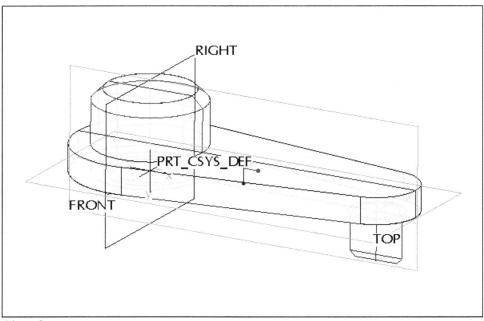

Chamfer

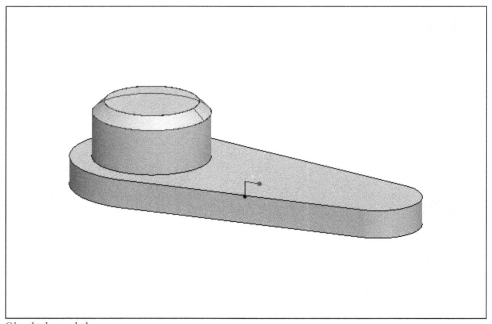

Shaded model

An extruded feature can also be used to remove material from the part instead of adding material to the part.

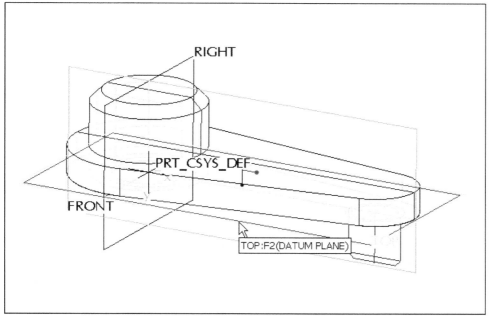

The sketch plane is selected

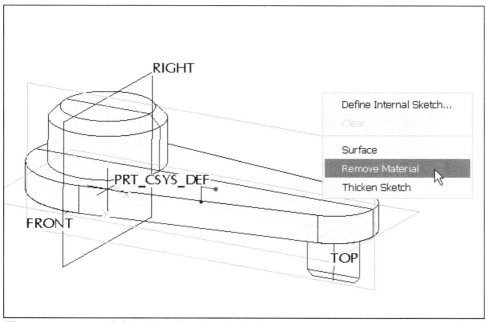

The remove material (cut) option is selected

The edge of the inner circle will drive the cut's diameter dimension.

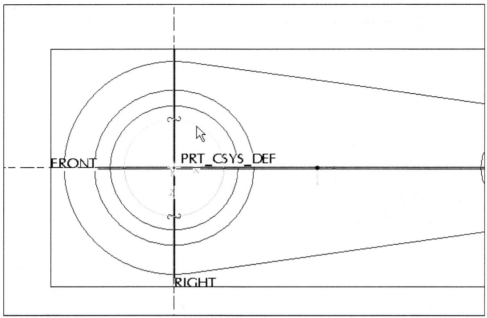

Inner circle is selected as the sketch geometry

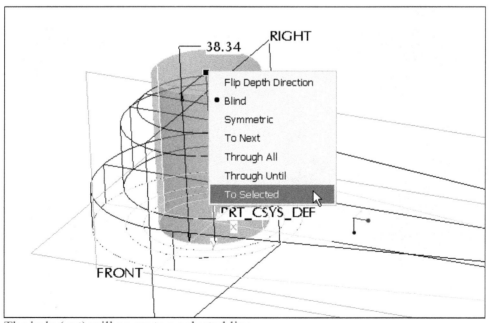

The hole (cut) will go up to a selected line

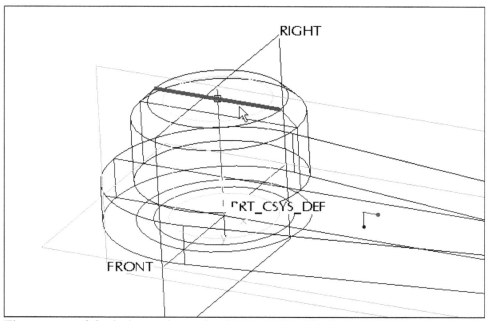

The extents of the hole are determined by selecting the imported geometry

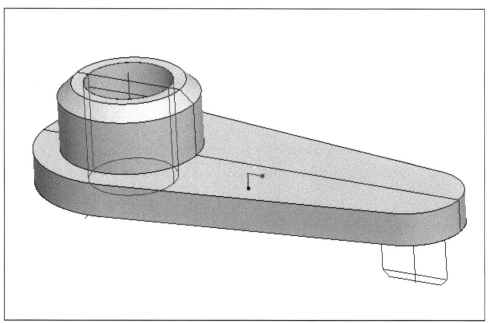

Hole is modeled

Revolving is another tool that allows any unique geometry to be rotated and extruded around an axis. You can rotate the geometry to however many degrees as required.

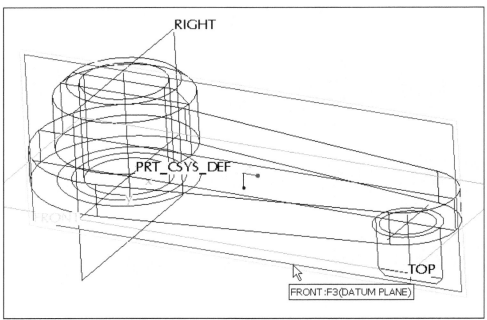

The FRONT datum plane will be used to establish the sketch geometry

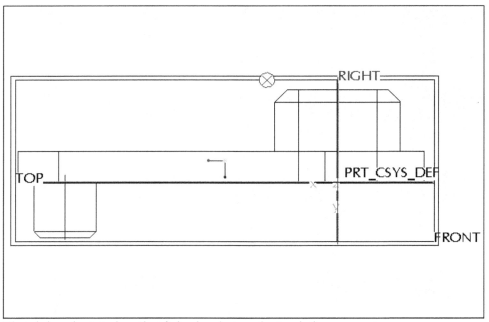

Sketch is orientated so that it is simple to select existing geometry

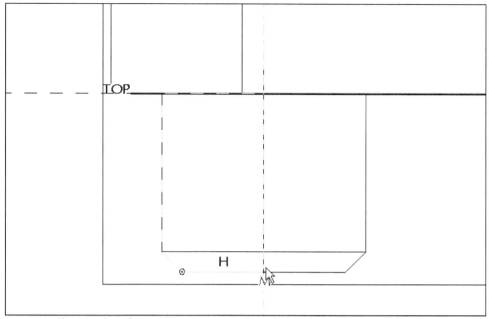

A centerline (axis of rotation) is created and the three lines of the feature are selected from the imported geometry

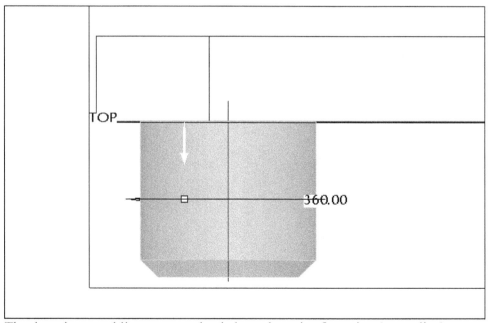

The three imported lines are revolved about the axis of rotation (centerline)

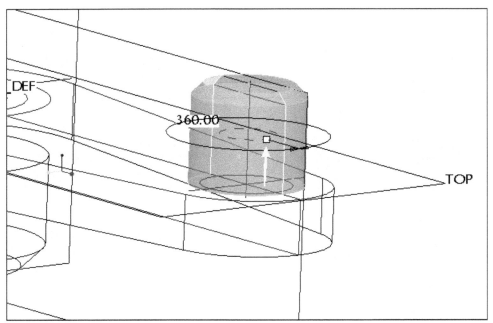

Previewed feature

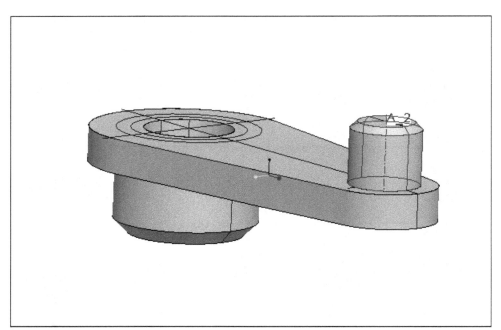

Completed model

With the exception of the chamfer, all solid features were created from and referenced to the 2D imported data.

A detail drawing can now be generated from the 3D model. The views, and dimensions are established, and the drawing format is selected as needed.

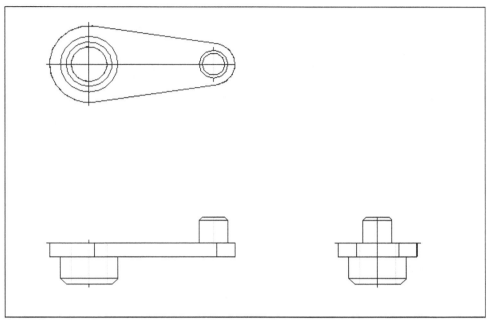

A 3-view drawing generated from the 3D model

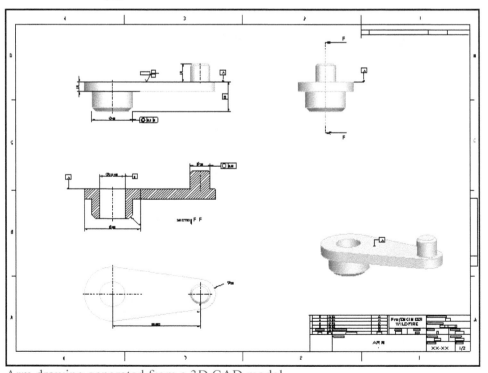

Arm drawing generated from a 3D CAD model

Accurate models & drawings reduce the cost associated with prototyping and manufacturing. Early discovery of design conflicts automatically reduce design time.

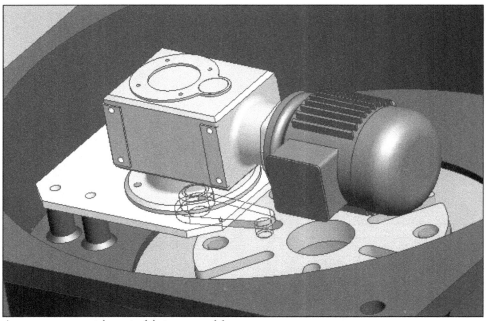
Arm component inserted into assembly

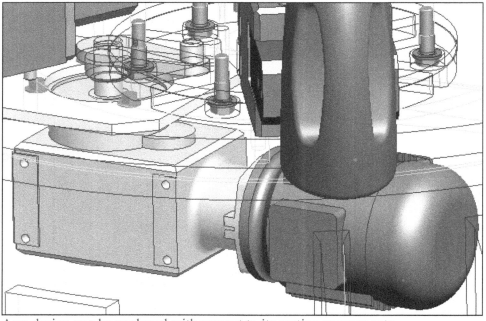
Arm design can be analyzed with respect to its mating components

By picking the packaged components, you can drag (moving the mouse/cursor back and forth) to show the mechanism movement.

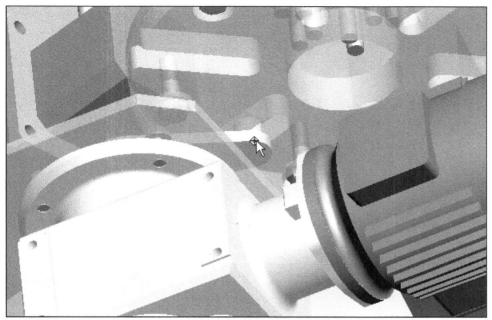

The component can be dragged to analyze its degree of freedom

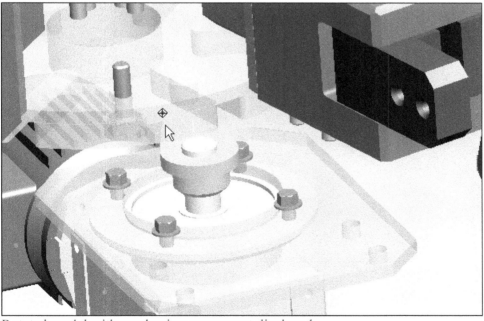

Rotated model with mechanism movement displayed

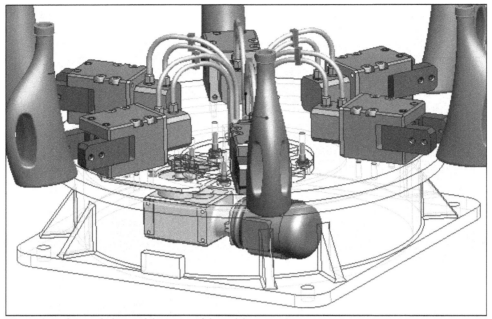

Completed revolving table assembly with new Arm component

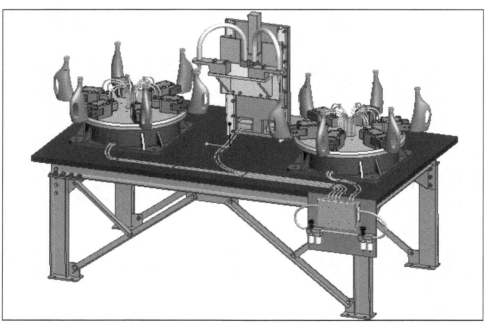

Complete assembly

Chapter Six Tools for converting 2D to 3D

- AutobuildZ from PTC
- Importing 2D drawings as DXF or IGES files
- Preparing a 2D drawing for use
- Defining orthographic views from imported data

A variety of software products are available for converting 2D data to 3D. In this chapter we will overview a product called AutobuildZ (pronounced as "Auto" + "builds"). AutobuildZ is a free plug-in application that includes a set of tools to create a 3D parametric, feature-based model from a 2D drawing, irrespective of the source (of the drawing). Within the drawing module of a 3D CAD tool, you can create a (2D) drawing by, importing a neutral file format such as IGES or DXF, or directly opening an AutoCAD DWG file. Additionally, you can sketch directly in the new (2D) drawing using draft entity creation tools and then use AutobuildZ to create 3D models.

AutobuildZ

With AutobuildZ, a plug-in application for Pro/ENGINEER Wildfire, design engineers have a comprehensive set of tools to create parametric, fully-featured 3D designs from 2D drawings by using a simple, wizard-based approach.

Feature wizards allow designers to select elements that define Pro/ENGINEER features directly from the drawing. Got extruded and revolved features needing valid section profiles? The fixing tools are right there in the feature wizards. And, creating fully detailed drawings from your new 3D design, with orthographic, section, detail and auxiliary views is a matter of clicks.

When we say easy, we mean easy.

Capabilities & Benefits:

- Use wizards to create Pro/ENGINEER features from 2D drawings
- Have constant visual feedback in 3D and 2D during feature creation
- Create automated drawings with associated views of the 3D design
- Select from multiple interactive tools for ease of use
- Rebuild designs from single, familiar drawing environment
- Increased efficiency from conversion of existing 2D data to 3D designs
- Training of current 2D workforce in 3D modeling concepts

AutobuildZ www.ptc.com/appserver/mkt/products/home.jsp

Note: An online tutorial is available at www.cad-resources.com that will guide you through the steps required to create the Arm using AutobuildZ. For online files, click www.cad-resources.com > **PTC Partnership** > **Chapter 6 Tutorial** > follow the instructions.

The tutorial (available from the web site) is designed to provide you with an overview of the suite of tools included in AutobuildZ. In addition, you be introduced to the complete process of building a parametric, feature-based model from a (2D) drawing using these tools. If you go through the tutorial, you will be guided through the steps involved in creating a 3D model from a 2D AutoCAD drawing. This tutorial assumes that you are somewhat familiar with the 3D CAD system used in the process. It is also assumes that you have successfully setup and installed AutobuildZ.

The AutoCAD drawing file (autobuildz_tutorial.dwg referenced in this tutorial) can be downloaded to your working directory. The 3D model that you will be creating should look similar to the image seen below.

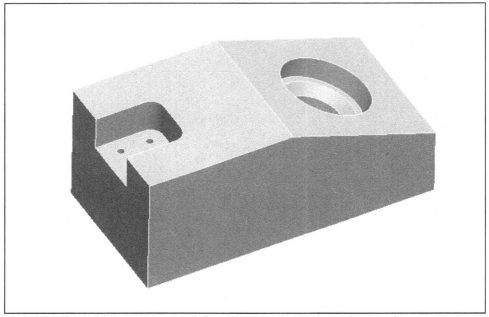

Completed 3D CAD part created from an imported 2D CAD drawing

This chapter provides a description and a set of illustrations of the AutobuildZ tool being used to create a simple part. The part is detailed in a drawing, which was imported from a 2D CAD source as a file.

Tools

AutobuildZ consists of the following set of tools:

- Clean Up Tools
- View Setup Tools
- Part Setup Tools
- Feature Creation Wizards
- Selection Tools

Workflow

The general process of creating a 3D model from the 2D drawing is: **Import 2D drawing > Cleanup drawing > Setup 2D views > Initialize 3D CAD parts > Create features > Save 3D CAD files**. Each of the steps in the process is described in detail here. You can complete the steps using the tutorial if desired.

The AutobuildZ demonstrations referenced below will allow to you see some of the techniques for conversion.

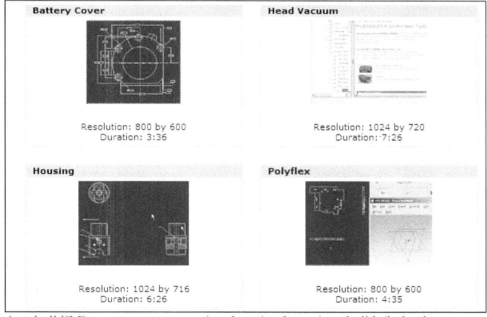

AutobuildZ Demos www.ptc.com/products/packages/autobuildz/index.htm

Importing 2D Drawings

Standalone 2D drawings are created by importing 2D data from a variety of sources (such as AutoCAD DWG or DXF) into your 3D CAD system. You can use the AutobuildZ set of tools on a 3D CAD system drawing in session to create 3D content.

Importing a 2D drawing in file format starts the process. DXF files are the de-facto neutral file formats for exchanging AutoCAD drawings or from other 2D sources. Once the file is imported, a drawing in session is initiated.

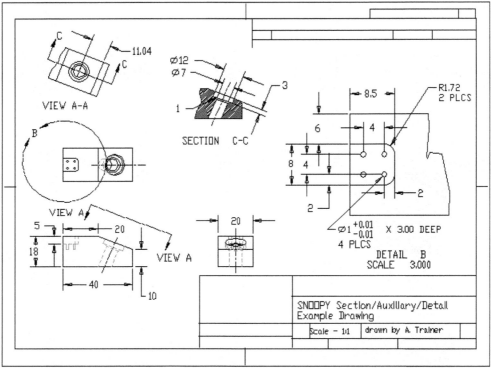

Imported file opened in a 3D CAD drawing

With a valid 3D CAD system drawing in session, the AutobuildZ set of tools is active and available for use.

Automatic cleanup of imported drawings

When a 2D drawing is imported into a 3D CAD system, it contains all the design content in the form of dimensions, notes, symbols and draft entities such as lines, arcs, circles, etc. that represent the geometry of the part. Some of these entities such as dimensions, notes, etc. are not used directly to define the feature so they are moved to layers that are then blanked.

The first step in preparing the drawing for feature creation is to cleanup the drawing. A combination of automatic and manual cleanup tools is used to clean up an imported drawing. During cleanup, unwanted entities such as dimensions, notes, symbols, etc, are collected and organized on separate layers. These layers are then blanked. At the end of cleanup, drawing sheets with only the draft entities that represent the geometry of the model are visible.

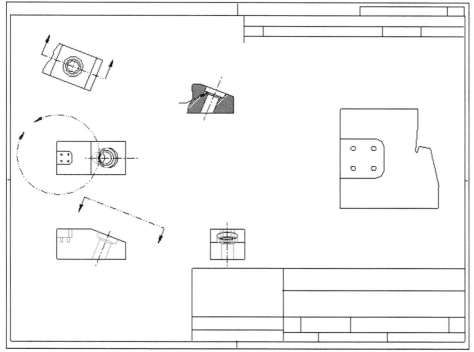

Automatically cleaned up drawing

The dimensions and notes are not displayed anymore on the drawing (layers are blanked). However, the drawing does still have a few entities (axes, centerlines, etc.) that don't represent the geometry of the model to recreate. Manually cleaning up the drawing is a step in the process of preparing it for feature creation.

Manual cleanup of imported drawings

Now that automatic cleanup of the imported drawing is completed, there are still some entities on the drawing that represent the leader lines, arrows, etc. that are displayed as lines and arcs that have not been blanked. These entities do not represent the geometry of the model. They can be manually cleaned up on the drawing by also moving these unneeded entities to other layers that are then blanked.

AutobuildZ has the ability to select all entities of the same color by selecting a representative entity in the graphics area.

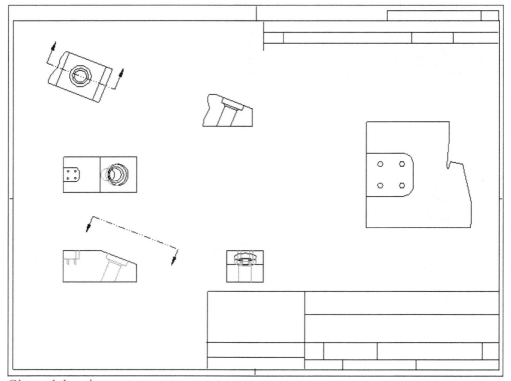

Cleaned drawing

You will now notice that only the entities that represent the geometry remain displayed in the graphics area. The next step to prepare the drawing for feature creation involves defining views on the drawing.

How to Define Orthographic Views – Compute the Drawing Scale

Drawing data imported to a 3D CAD system may or may not contain views. Drafting entities of the imported drawing can be selected and logically grouped to represent views in AutobuildZ. Orthographic, detail, section and auxiliary views using the AutobuildZ view setup tools can be created. When defining orthographic views, the drawing scale can be calculated from reference dimensions and draft entities on the drawing or default drawing scale (1:1) can be accepted. As a designer, you will note that there are three logical groups of entities on the drawing representing the orthographic views (TOP, FRONT and RIGHT). Since, this is a drawing in third-angle projection; it can be assumed that the group of entities in the bottom left represents the front view. It can also be assumed that the group of entities to the right of the front view is the right view and the group of entities above the front view is the top view.

The FRONT orthographic view is the default view selected. The drawing scale is then established. The reference dimension and the reference entities are selected to establish a reference scale. The drawing scale is calculated and displayed. This scale overrides the default value for the orthographic views in the setup.

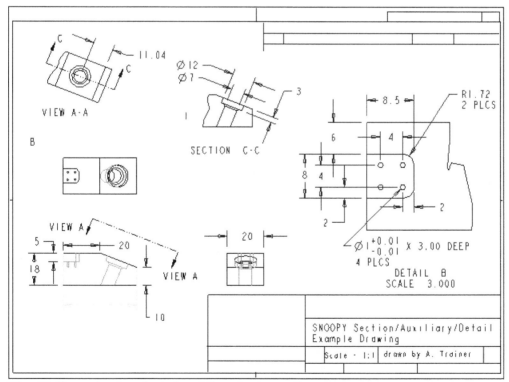

The **40** dimension is selected and then the two references

125

How to Define Orthographic Views – Selecting Entities for each View

The Front view is established by selecting the desired entities for the view. Once the orthographic view is created, there is a border around the entities selected for the view.

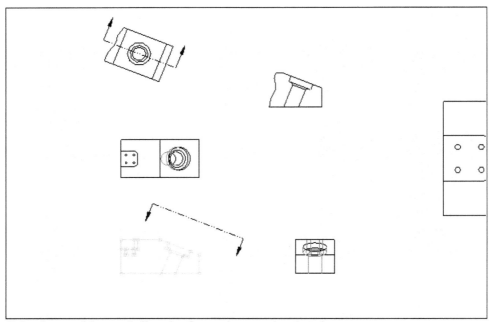

Selecting entities to define the front view on the drawing

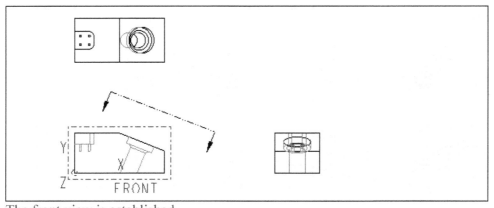

The front view is established

The TOP and RIGHT views on the drawing can be established using the same set of steps.

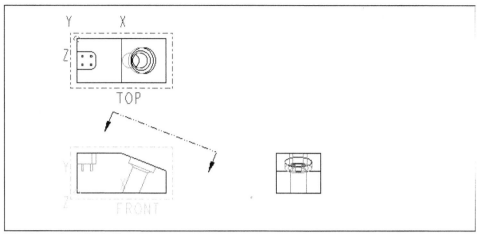

The top view is defined

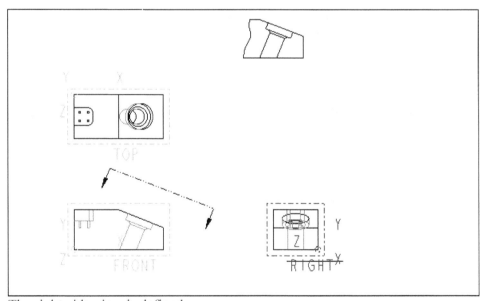

The right-side view is defined

Initializing a New 3D CAD system Part

One of the goals of using AutobuildZ is to create a 3D parametric, feature-based model from the 2D drawing. A 3D part needs to be set up in session where features are created. Once the part is initialized, it gets "associated" with the drawing (the part refers to model as the "Drawing Model" in the context of the drawing). The part setup tools of the 3D CAD system enable users to initialize one or more 3D CAD system parts using a template start part. There are default templates (for inch and millimeter units) provided with AutobuildZ. The appropriate template is selected based on the units of the drawing. These template parts include the TOP, FRONT, and RIGHT default datum planes. These planes will be automatically mapped to the (orthographic) views of the same names defined on the drawing during view setup. When the model is initialized, AutobuildZ also makes this model as the "active" model in which features are created.

Creating 3D CAD system Features

Using AutobuildZ feature creation tools, all the components can be selected that define 3D CAD system features such as extrusions, revolves, holes and even datum features directly from the drawing. Feature creation wizards in AutobuildZ are provided with an integrated environment to define and create 3D CAD system features. A series of solid 3D CAD system features that define the 3D model are created next. An extruded feature (base feature) is created first.

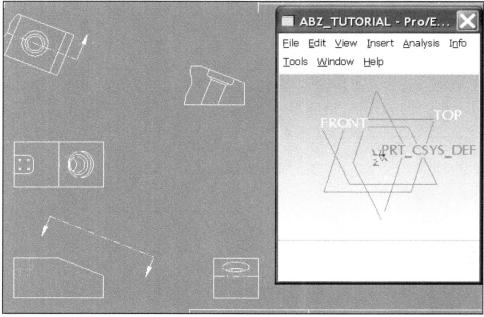

Creating the base feature

The section profile for the extrusion to be created is selected next. By selecting one of the entities in the FRONT view as shown below; the entire chain of entities is selected forming a closed loop. The section profile selected is automatically validated to make sure that the section is valid for the feature being created. In this case, the section profile is a single closed loop and is valid.

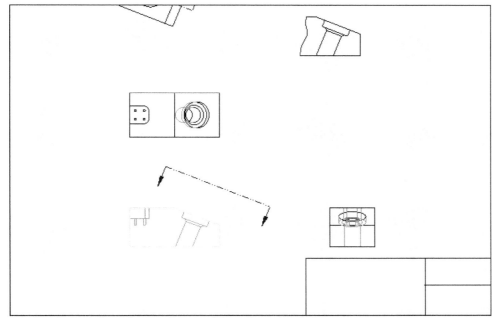

Interactively select the section profile to be extruded

A 2D line entity in another view is selected to represent the sketching plane as shown below. AutobuildZ will automatically find either an existing surface or datum plane to use as the sketching plane. In this case, the FRONT datum plane is determined to be the sketching plane based on the entity selected.

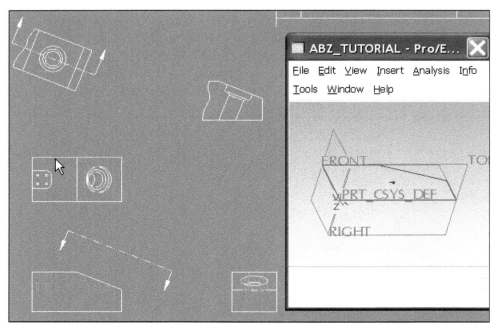

Select the sketching plane reference

Now the extrusion depth must be established by selecting a 2D line entity that represents the depth. Since, the entity selected is perpendicular to the reference entity for the sketch plane; AutobuildZ will try and find an "up to surface" as the extent of extrusion. However, since no surface exists, a blind depth is calculated and displayed.

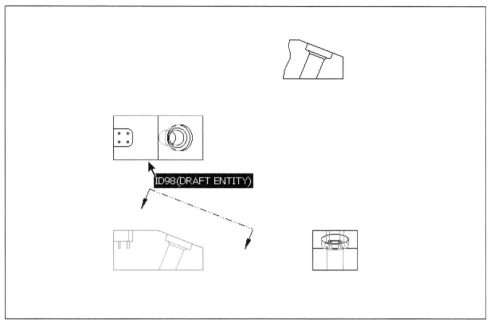

Defining the depth of the protrusion

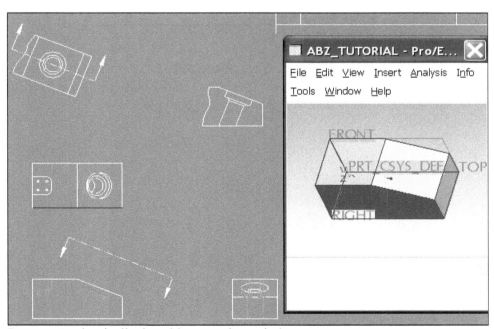

Base protrusion is displayed in a preview window

Creating additional 3D CAD system Features

Now that the base protrusion is complete, additional features will be added to the model. Continue the 3D model building process by creating the extruded cut. The section profile for the cut being created is determined by the selection of the three lines and two fillets (loop).

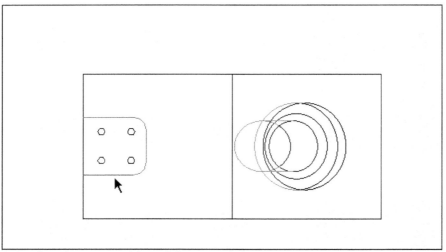

Selecting the type entities of the extruded cut

The entities selected do not need to form a close loop. This is because AutobuildZ detects that there is an existing entity (edge) that might be used to close the section profile once we have selected the sketch plane.

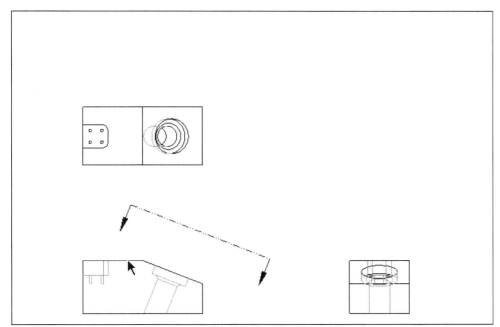

Select the reference entity for the sketching plane

The sketching plane is defined next. Select the entity as shown below. In this case, an existing surface (top surface of the base feature) in the model is determined to be the sketching plane.

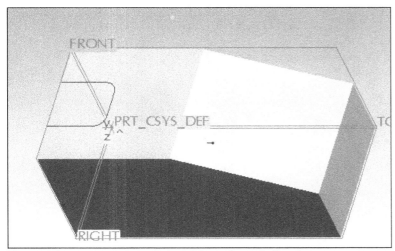

Cut section geometry and sketching plane defined

133

Now the depth of the extruded cut feature must be defined. A 2D line entity that represents the depth is selected as shown below. The cut feature is created and can be displayed in the preview window.

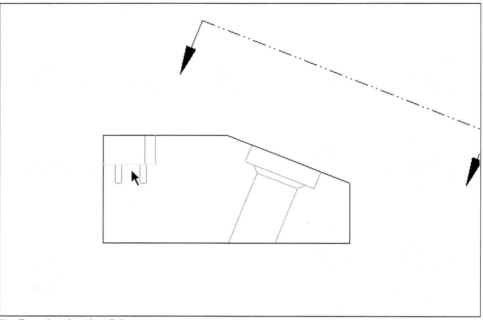

Define the depth of the cut

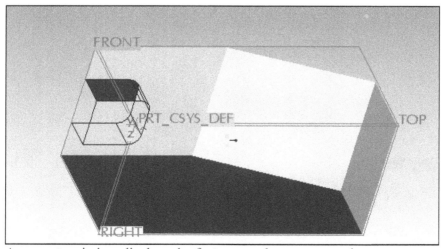

A separate window displays the features as they are created

Create a Hole

The next step is to create additional features in the 3D model. There is a set of four holes on the bottom surface of the cut feature. The hole feature creation wizard will be used. The circle entity that represents the hole being created is selected to define the diameter of the hole.

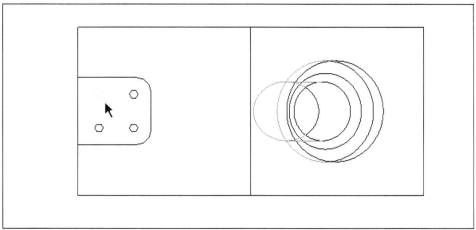

Define the diameter of the hole by selecting the circle entity

The surface that the hole lies on is established by selecting the edge line of the sketching plane, in this case, an existing surface (bottom surface of the cut).

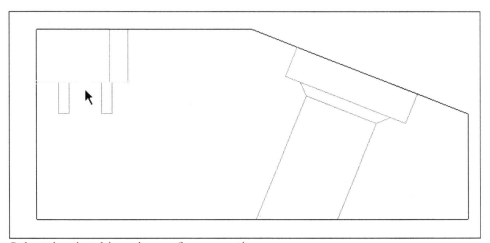

Select the sketching plane reference entity

Now the depth of the hole is required. Similar to defining the depth of the base protrusion, a 2D line entity that represents the depth of the hole is selected as shown below. The depth of the hole is calculated and the hole feature is created.

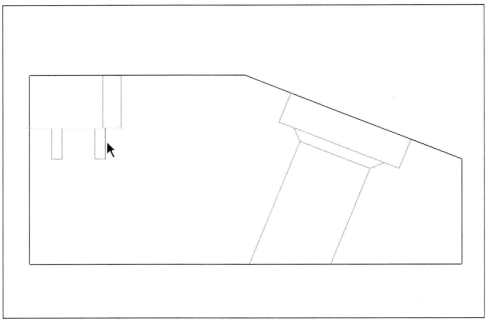

Define the reference entity for the depth of the hole

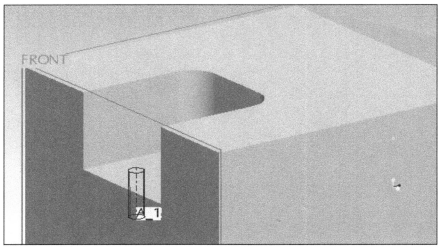

Previewed hole

Now that the first hole has been defined, the remaining holes are created using the same process.

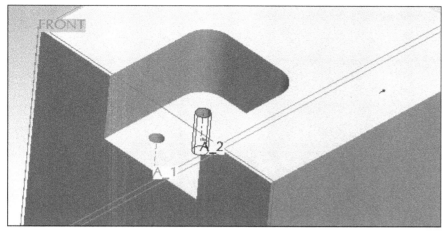

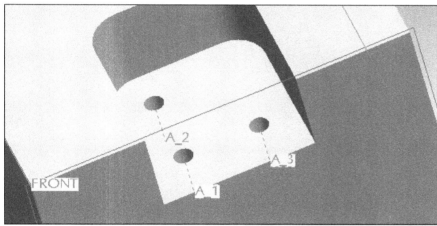

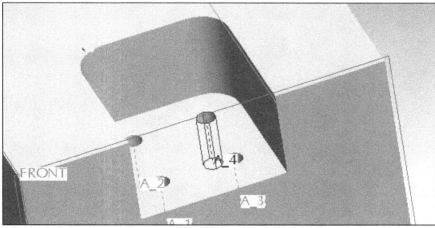

Four holes created using AutobuildZ

A pattern tool, when modeling in 3D part mode, could also be used to create the other hole features.

The final feature can be created using a standard hole tool of the 3D CAD system.

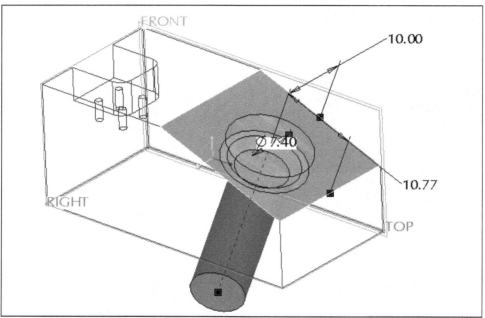

3D CAD Standard hole tool is used to model the last feature

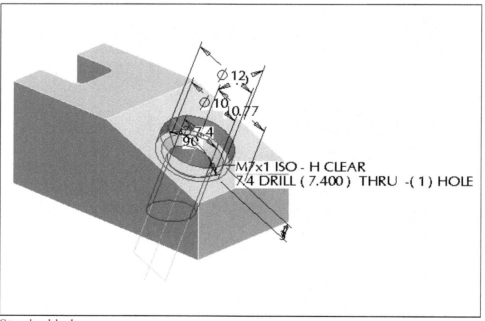

Standard hole

The current drawing being worked on can be saved at any time. A part model, that is associated with the drawing (i.e. the part where the features are created), will also be saved to disk anytime the drawing is saved. This will save the drawing and any associated parts to the current working directory.

What is important is that when you use AutobuildZ tools on a drawing, specific information about the AutobuildZ process (for example; the views definition, layers, etc.) also get saved within the drawing automatically. This is convenient because the entire conversion process is initiated in a single session. Modeling can stop at any time and then resume the process later where it was left off.

Since, the part model is associated with the drawing; it will be retrieved when the drawing is retrieved.

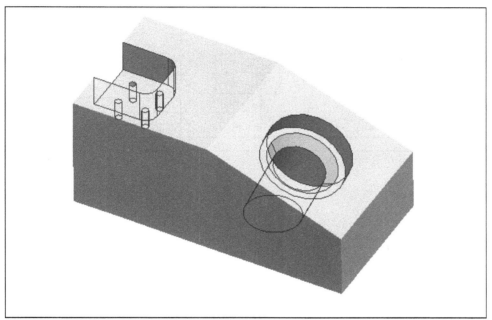

Completed design

This completes an overview of the basic steps in the process of creating 3D CAD features from 2D legacy data using AutobuildZ.

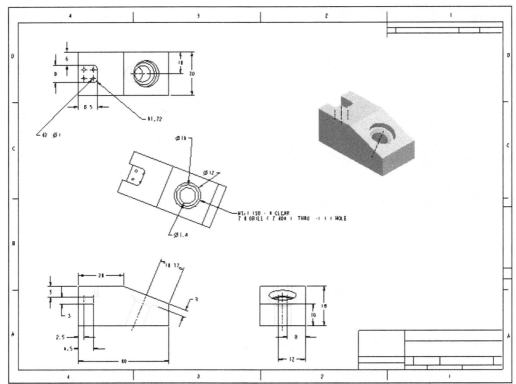

Drawing generated from 3D model

Chapter Seven **Your workforce**

- Developing your workforce
- Creating your company training plan
- Online versus in-class training possibilities
- Existing educational opportunities
- Contributions from value-added resellers and consultants

Your workforce is of course the bedrock of your design department. Software and hardware do not do the work. People do the work using the tools they are happy with and with which they are comfortable. Neglecting to understand this will cost you the ability to evolve your design and engineering department into a 3D savvy operation, and possibly cost you your job in the process. If the cost in terms of dollars is not understood and if the loss of productivity due to a poor transition plan is not appreciated, it could spell doom for your transition.

This chapter describes various educational opportunities and training methods. You and your transition team should explore and analyze the possibilities and then create a plan to follow for your move from 2D to 3D CAD.

The foundation of a successful move from 2D to 3D CAD is training. Training is not a one time, one week, get up and running task. It is a continual process involving the commitment of the company and the involvement of the full design and manufacturing departments. Making available 3 to 4 mechanisms for learning and keeping up-to-date will insure a well-trained and high quality workforce who will be and stay productive. Skimp here and you lose in every area of design, manufacturing, quality and efficiency gained by introducing a high level 3D CAD system. CAD managers and systems people should also be required to learn the 3D CAD tool and get periodic training regardless of their role in the design and documentation process.

Getting management on board early in the process will ensure its success. Costs for retraining the existing workforce must be understood from the start. Commitment up front for transition training and for work-life-long training is extremely important. If your existing workforce sees management waver, or not supply the proper tools and support, they will resist and possibly sabotage any new software system implementation. Industry is rife with examples of how an existing and possibly aging design department can undermine new technology and software from being successfully adopted.

Every product designer knows that designing in 3D solid modeling requires a very different approach from that of 2D CAD design. To ease this transition, the 3D software itself should help wherever possible, it should be simple to use and understand. It should be modern, and familiar, like other desktop applications that users are comfortable with. It should also make extensive use of automated training and tutorials.

Hiring new people will be easier if you have a mainstream, popular, high-end 3D CAD system as your design tool. Few 3D designers would ever take a job that required the use of a 2D CAD system. I have had literally hundreds of students (my students are heavily degreed and have an average of 7 years experience) who have refused jobs using a 2D CAD system, even those who knew the 2D CAD system well, and even during one of the cyclical downturns seen in Silicon Valley. It would limit their future career choices and prospects and lock them into a limited and possibly dying technology. "It's a step backwards in technology". In other words, having a high quality 3D CAD system will actually attract new talent.

Gone are the days of incredibly powerful, but equally complicated 3D CAD systems that only the most advanced users could master. Microsoft's Windows operating system has been adopted as an engineering standard; more individuals than ever before are interested in using 3D CAD solutions; and no one has the tolerance for spending weeks on end trying to learn overly complex software. The best 3D CAD tools provide powerful capabilities, in a scalable, easy-to-learn and easy-to-use package, that's instantly familiar, and allows designers and engineers to spend their time delivering great products, instead of learning how to use the tools. Furthermore, the tools must feature comprehensive, built-in tutorials and online training options that allow users to learn various topics at their own pace.

Look for a 3D CAD system that also has an extensive partner network offering instructor-led training in a wide range of design fields. Get your workforce on board and enthused by doing your homework and giving them the proper extended training. Existing personnel will see the managers and the company's commitment to providing assistance as proof of their support throughout the transition, and beyond.

I have taught drafting, design, and engineering graphics in junior high (JH), high school (HS), technical schools, community colleges, private engineering colleges and at universities. I have also given week-long industry training courses on 3D CAD and had the opportunity to assist companies in moving to an integrated 3D database for manufacturing. This experience has given me a unique view of the whole arc of engineering design graphics education and "industry targeted" training. What I have learned during these experiences is that the higher the quality of education/training, the higher success rate of the students/employees and subsequently the benefits to their employers.

CAD Company training services

Every CAD Software Company offers a complete and comprehensive set of tools to learn and stay current with their software.

Training Services

PTC University and Learning Resources

Boost productivity with PTC University! Whether you're looking for a specific training course, ways to improve the proficiency of experienced users, or a personalized corporate training program to meet your specific training time and budget, PTC University can help.

- Search the PTC University Course Schedule to find a class to suit your needs
- Consult PTC Role-Based Learning Paths to find the course that's right for you
- Check out PTC Training Courses in live classroom, virtual classroom, or Web-based training formats
- Subscribe to PTC University eLearning for 24/7 access to over 500 hours of Web-based content
- Enroll in a Precision Learning Program and improve team proficiency by 20% -- guaranteed!
- Other Training Questions
- To purchase an eLearning subscription or for help enrolling in a class, contact PTC University

P T C University

PTC Training Services www.ptc.com

A live instructor in an actual classroom presents live classes. The student/employee will also have access to the 3D CAD software so that they can practice what they are learning. The down side to this training is that the employee has to travel to the classroom, which may be in another city. There is generally a lot of material to cover and the pace of the class is swift. Sending *entry-level users* to these classes may not be the best use of your resources. Employees with mid-level or advanced skills will benefit the most from short intensive trainings that target areas where they are deficient.

Virtual classes offer a new, cost-effective way to attend live training without leaving your home or place of employment. Virtual classes are taught by a live instructor through the Web allowing students to experience the same course content and interaction found in live classroom training.

Webcasts on specialized subjects will also provide your staff with the ability to view and to participate in demonstrations and question and answer sessions. Web-based training allows you access to the same material as in the virtual class, but there is no instructor to answer your questions. This type of training can be utilized as a self-paced course performed during work hours, after work hours, or at the employee's home provided that they have access to the 3D CAD software.

VARS (value added resellers) and Consultants

A VAR is an independent company that provides assistance in combining the correct software and hardware solution for your business. A consultant is an independent person or firm that provides specialized support for design and for system integration.

Consultants can be expensive for a small company to employ over a long period of time, but do add needed assistance when specific needs are identified. Unlike VARS, which are tied to the software companies as partners, consultants are independent and may provide more objective analysis and assistance. Regardless, both are in the business to make money. Your transition team needs to properly identify your needs and the budget constraints within which you are working.

In many cases, your local or regional VAR will be your point of contact for training and other support. Partnering with a competent VAR will enhance your chances of success. Remember, it is your company's workforce requirements that are the driving force behind these partnerships. Be specific in what you need and do not be pressured to buy training that is outside the requirements of your companies design focus.

Resellers

PTC's Global Network of Resources

PTC Channel Advantage is an industry leading program developed with our worldwide Value Added Reseller community to provide PTC products, services, training and support to our small and medium size business customers.

- Find a Reseller Partner
- Contact Us Directly
- Become a Reseller
- Partner & Reseller Portals

This program allows for our Reseller Partners to become trained and certified in our various product and solution implementation methodologies, and has three distinct levels. PTC Channel Advantage Silver, PTC Channel Advantage Gold and PTC Channel Advantage Platinum. These stages are based on increasing levels of investement requirements and associated program benefits.

VARS www.ptc.com/partners/var_channel.htm

Online Training

On-line, computer-based training (CBT) is simpler and cheaper than traditional classroom training, and can be accessed 24x7 from any computer, anywhere. Running concurrently with a live 3D CAD session, a high quality CBT uses interactive examples, hands-on exercises, and detailed reference materials to create real and lasting (learned) knowledge.

At our college, we have used the CBT CADTRAIN's COACH for the last 12 years. CADTRAIN has provided materials that supplement our existing books, tutorials, lectures and projects.

Purchasing a floating license of an online CBT training package will provide your workforce with continual access to self-training. When new versions of the 3D CAD software are released, individuals in your workforce can activate the update training module and in a few hours be familiar with the new interface and enhancements that are available with the new software version. As software changes from version to version, a CBT allows the user to target aspects of the software change that affects their job. New 3D CAD users can ramp-up their skills, and experienced users can brush-up on rarely used features. Keeping skills current is easy with the detailed update training modules, issued for each new release of the software.

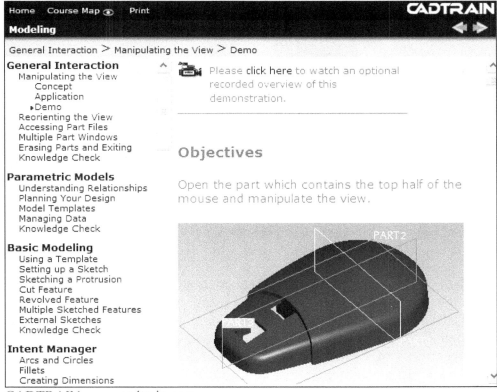

CADTRAIN www.cadtrain.com

University "trained" engineers

If you are in the market for newly minted engineers to grow your workforce, plan on spending time and money retraining these beginners.

In the last few years, the college where I teach is seeing a new "feeder" group coming into the program. Instead of high school graduates, we are seeing university graduates. In general, we train/teach existing engineers, designers and drafters who on average have 7 years of experience and multiple degrees. Before 1990, we had 80% of our feeders from the local high school drafting programs. Now, only about 10% of our program caters to recent high school graduates. But, we now get some recent college graduates who cannot land a job. This is a new feeder stream not seen previously.

Mechanical Engineering at California Polytechnic State University
ceng.calpoly.edu

Local engineering school graduates have been showing up frustrated that no one will hire them because they cannot "do anything". They cannot design with any of the major 3D CAD systems used in local industry. Many of them take over a year to get their first job. Others, may get a job, but must be immediately trained in 3D CAD at a high cost to the company. The sad thing is that CAD companies try and sell industry on the concept of bringing any new engineer up to speed on their 3D CAD system with a week of training. *Do not believe this.* Any engineer, regardless of their IQ and in many cases in spite of it cannot be productive let alone master a complex 3D CAD system in less than six months. This six month period can include an expensive week of instructor lead training.

External trainings must be supplemented with web based tutorials, web casts, CBT training available on their workstation and home computer, and in-house tutoring. This will ensure that the person actually learns how to create their designs with the proper design intent.

Poorly trained engineers create models that fall apart with even minimum changes. Also, without work life-long training your workforce will not be able to keep up with the 12-month cycle of software changes that every CAD adheres to. This may sound complicated, but it really only requires that you and your company partner with your workforce to provide continual support and resources to keep education and training as the bedrock of your engineering and design staff. A happy workforce is derived from not expecting miracles after a week of training, but a balanced program of training, education, tutorial resources, and CBT, etc.

One young student had recently graduated from a major northern California university as a mechanical engineer. The student took one engineering graphics class, which used BozoCAD. As the student stated, BozoCAD is not used anywhere in local industry, so how was the student to get a job (besides, 40 hours on BozoCAD that was three years out-of-date was next to useless). After six months taking classes of appropriate 3D CAD programs at our facility, he is now employed. Sadly, most engineering professors' at large universities do not know or do not want to learn and relearn complex 3D CAD packages.

Only university programs which emphasize the technical and design aspects of engineering actually teach 3D CAD to a level that is usable and even they are so overloaded with classes in math, physics, chemistry, engineering, etc. that engineering graphical design warrants at most, two classes. And these engineering graphics classes are normally taught in the sophomore year. By the time the engineering student graduates from a 4 or 5 year program, the software they used has gone through 3-4 version changes and the student barely remembers the command structure.

Locally, we have seen many of the local university engineering programs send their engineering students to us for their CAD learning. We gladly extend a flexible schedule to them so that our classes fit within their complex and taxing engineering program.

Though what has been written here may seem unduly harsh and critical of university engineering programs, it is an accurate reflection of the current state of engineering education. There are many at the university-level who are pursuing reform and change at their schools and with state engineering education committees. There are also many universities that have seen the writing on the wall and have added excellent CAD graphics classes to their curriculum. In order to make their students more employable, these universities are attempting to integrate CAD training into the traditional engineering curriculum.

Mechanical Engineering University of British Columbia www.mech.ubc.ca

Throughout the nation and in many countries around the world, engineering education has been evolving to include a heavy dose of CAD. Regardless, as an employer looking for new engineers trained and ready to be productive, you will not find many appropriate applicants. When you hire graduating engineers, expect to spend time and money getting them up to speed on your CAD system. In fact, having a continuing training and education program in place will attract the best and brightest.

A list of university engineering programs with integrated CAD classes is provided on the Transition Resources page of the web site www.cad-resources.com.

Public Community College and Technical Schools

Since I am a community college instructor and have been for a majority of my life, I of course will be prejudiced in favor of these programs. Community College (CC) classes usually offer an excellent value if a local school has the software you want taught. If your area has a number of industries involved in design and engineering, you may consider partnering with them to set up community college CAD classes that teach the software that you use or plan to migrate to in the future. Software companies have made their software available by establishing educational programs (PTC's Education Program).

De Anza College's CAD facility www.deanza.edu/cdi/

CC's are normally less expensive than any other educational methods available. With a CC, you can get the college to provide the facility and hardware for training. This also allows your employees to get college credit and, certificates, and even an AA or AS degree for their efforts.

As long as the students are available, a CC will be interested in a new or supplemental program. Your company may also supply the instructor. In other words, the school would pay for the instructor. The facility will be free, therefore not tying up your company's computers and office space. The cost of the training will be limited to your company reimbursing the employee for registration, class units, and book, if your company has such a program in place. Your employees will have 75-100 hours learning the CAD program per class they take. Their learning curve will be excellent.

In our program, a majority of the CAD classes are available evenings and Saturdays making them convenient for working students. In the summer we run 3-week morning classes for companies that can spare releasing their workers for an extended time. Many community colleges offer similar comprehensive programs. Though I have referenced my own program here, there are thousands of worthy college CAD programs throughout the world.

The program created at our college is an example of a comprehensive and constantly evolving CAD department. Over the last 25 years, we have added and eliminated CAD programs as the local industry and job requirements have changed. The CAD program at De Anza College officially started in 1984 with the introduction of Computervison to the curriculum. Previously, (1965-1984) the Drafting department had a traditional drafting and design certificate and degree program. During the last 25 years, the CAD program has changed names and divisions a number of times while continuously evolving and reinventing itself to meet the needs of the high-tech job market in Silicon Valley. In the past, Computervison, Personal Designer, Calma, and CADAM were offered. Presently, AutoDesk's AutoCAD and Inventor, 3D Civil, and Architectural Desktop, PTC's Pro/ENGINEER, SolidWorks, and Unigraphics are the CAD design packages taught in the program.

Certificate of Achievement - Autodesk

General Requirements
1. A minimum 2.0 grade point average in these units.
2. A maximum of 6 quarter units may be transferred from other institutions.
3. See Counseling Center for academic planning and to apply for certificate.

Course Requirements		Units
CDI 80B	AutoCAD (Beginning)	4
CDI 81B	AutoCAD (Intermediate)	4
CDI 85B	AutoDesk Inventor	4
	Total Units Required	12

Certificate of Achievement - Pro/ENGINEER

General Requirements
1. A minimum 2.0 grade point average in these units.
2. A maximum of 6 quarter units may be transferred from other institutions.
3. See Counseling Center for academic planning and to apply for certificate.

Course Requirements		Units
CDI 70B	Pro/ENGINEER Wildfire 3.0 (Beginning)	4
CDI 71B	Pro/ENGINEER Wildfire 3.0 (Intermediate)	4
CDI 72B	Pro/ENGINEER Wildfire 3.0 (Advanced)	4

Complete 8 units from the courses listed below:

CDI 73B	Pro/ENGINEER Wildfire 3.0 (Pro/SHEETMETAL) (4)	
CDI 74B	Pro/ENGINEER Wildfire 3.0 (Pro/SURFACE) (4)	
CDI 75B	Pro/ENGINEER Wildfire 3.0 (Pro/MOLD) (4)	
CDI 76B	Pro/ENGINEER Wildfire 3.0 (Pro/CABLE) (4)	
CDI 77B	Pro/ENGINEER Wildfire 3.0 (Pro/MECHANICA) (4)	
CDI 79B	Pro/ENGINEER Wildfire 3.0 (Pro/Update) (4)	
	Total Units Required	20

Certificate of Achievement - SolidWorks

General Requirements
1. A minimum 2.0 grade point average in these units.
2. A maximum of 6 quarter units may be transferred from other institutions.
3. See Counseling Center for academic planning and to apply for certificate.

Course Requirements		Units
CDI 60B	SolidWorks (Beginning)	4
CDI 61B	SolidWorks (Intermediate)	4
CDI 62B	SolidWorks (Advanced)	4
	Total Units Required	12

Certificate of Proficiency - Computer Aided Design - Mechanical

General Requirements
1. A minimum 2.0 grade point average in these units.
2. A maximum of 18 quarter units may be transferred from other institutions.
3. Demonstrated proficiency in mathematics and English.
4. See Counseling Center for academic planning and to apply for certificate.

Complete two of the options A-D listed below, plus additional CAD courses to a minimum of 28 units.

Course Requirements Options	Units
A. Certificate of Achievement - AutoDesk (12)	
B. Certificate of Achievement - Pro/ENGINEER (20)	
C. Certificate of Achievement - SolidWorks (12)	
D. CDI 57A Simultaneous Product Development (4)	
Total Units Required	28-32

A.S. Degree - Computer Aided Design - Mechanical

A.A./A.S. Degree Requirements
1. Completion of a minimum of 90 quarter units of college credit, including:
   • A minimum of 24 quarter units must be earned at De Anza College.
   • A maximum of 22 quarter units from another institution may be applied toward the MAJOR.
2. Completion of all General Education requirements (31-42 quarter units) for the A.A./A.S. degree with a minimum 2.0 ("C") grade point average.
3. Completion of all Major requirements with a minimum 2.0 ("C") grade point average. Major requirement courses can ALSO be used to satisfy G.E. requirements.
4. Completion of all De Anza courses must be with a minimum 2.0 ("C") grade point average, and all De Anza courses combined with courses transferred from other colleges or universities must be with minimum 2.0 ("C") grade point average.
Students are encouraged to see a counselor or adviser for academic planning and to apply for the degree.

Major Requirements - Complete all course requirements for the following CAD Certificates of Achievement: AutoDesk, Pro/ENGINEER, and SolidWorks, plus the following:

CDI 51	Geometric Dimensioning and Tolerancing	2
	Total Units Required	46

At De Anza College, we have a very comprehensive CAD program. Most community colleges will only have one or two CAD systems available.

For a list of CC's offering CAD programs, click: www.cad-resources.com **> PTC Partnership > Transition Resources > Colleges**.

Local K-12 Programs

For long term planning for your future workforce look to the JH and HS programs in your area. Believe it or not, young computer-savvy students will someday be future employees. As a long-term strategy for your employment requirements, setting up an internship (HS summer, or college) would also help build a future workforce. Oddly, some of these young students could be more comfortable with a 3D CAD software package than your existing professional workforce. See the Transition Resources page on the CAD-Resources web site for a list of schools with excellent CAD programs.

Many 3D CAD companies have setup free software grants to JH and HS. As an example, PTC's Pro/ENGINEER Schools Edition is available free to all secondary schools. An example of the type of work accomplished in school programs can be seen with the designs from Gloucester High School, MA. No comment should be necessary as I would hope that these images would sufficiently impress.

Gloucester High School, MA www.ptc.com/for/education/gallery/ghs.htm

In many cases, HS graduates entering college programs have only been exposed to 2D CAD. Free 3D CAD schools edition software grants should start to bridge the educational CAD gap created by high schools only offering 2D CAD. As a student moves from HS to college, to the workplace, they will bring with them a love for the "first" CAD system they see and master. If someone uses a particular CAD system in school, you will look for a job that has that CAD tool as the design system used internally, or encourage the company to switch.

Student and Tryout Editions (Software)

For your business, investing in expensive instructor lead week-long trainings can be successful if the employee has access to the new software at home. Taking intense 35-hour training, sometimes even sharing a computer, etc. will only frustrate the employee if they do not have the ability to go back to work or home, and practice the methods and concepts they "learned" during the training.

Having a set of manuals available and a CBT module such as CADTRAIN will also ensure that your investment in their skill development is successful. For your employee to become truly productive on any 3D CAD system they will need continual support and access to a wide range of resources.

Many CAD software packages allow employees to "borrow" a license to use on their laptop, or home computer for short periods of time.

The availability of software in the form of a free **Tryout Edition** or a low cost **Student Edition** [$100.00-$250.00 (US)] is essential for success in many instances. Having a home version of the cad package will enable them to push their understanding of the software to its limits in their spare time, outside the classroom. Get your employees a copy of the Student Edition and a set of tutorials and books so that they can use it at home.

PTC Pro/ENGINEER Wildfire 3.0 SE & Lamit Text for Academic Use

 for Microsoft Windows

Academic Proof Required

Item Number: 30144903

Students Editions from JourneyEd www.journeyed.com

For Pro/ENGINEER, my textbook comes bundled with a copy of the Student Edition software. The textbook and Student Edition software can also be purchased separately. Almost all 3D CAD systems have this type of tutorial-software combination available.

Other Training materials, publications and the web

The CAD program, at the college that I currently teach at, utilizes a variety of books, training manuals, tutorials and other printed materials. Reasonably priced books from SDC are used for classes in five different CAD software packages. Reputable books and manuals are also found at CADQUEST www.cadquest.com, and major publishers such as Thomson (ITP) Engineering.

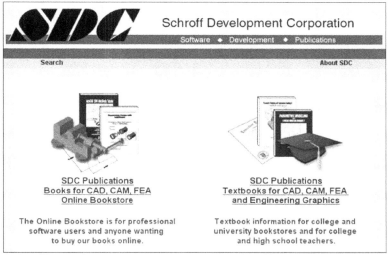

Schroff Development Company www.schroff.com

Frotime has also been used at this facility with success. This company provides training and tutorials covering a wide range of CAD topics for Pro/ENGINEER. Other 3D CAD companies have similar external educational materials available.

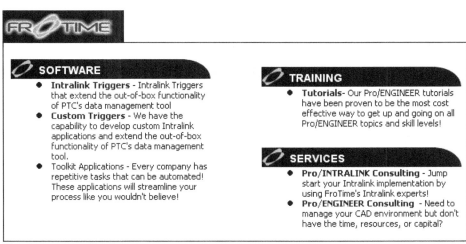

Frotime www.frotime.com

User Groups

User groups can be a great asset in the identification of local talent, provide assets and assistance for your move to 3D CAD, and provide support for individuals in your workforce. In areas where there is no local user group for the 3D CAD system of your choice, it is very simple to start your own. Many companies support local user groups by sponsoring events and providing meeting space. Every 3D CAD company has materials available to guide a new group from creation to their first meeting. User group kits can be downloaded directly from the software provider's web site.

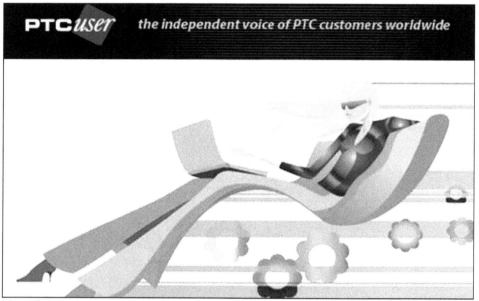

User Groups www.ptcuser.org

The Customer Care Zone
A PTC Product Assistance Web Site

Pro/ENGINEER Regional User Group Information

The Pro/USER group Technical Committees (TCs) provide another, more interactive, forum for providing information to PTC PLMs regarding product direction and has been a very successful program, with approximately 80% of member suggestions appearing as product enhancements. It is the independent voice for users of software produced by PTC. Pro/USER is the sole entity recognized by PTC as the official worldwide representative of its customers and is an independent, not-for-profit corporation, organized by volunteers from companies owning PTC products. We are self-funded, our revenues coming from our annual conference, magazine advertising and sponsorship from our Industry Partners.

User Groups www.ptc.com/carezone/user_groups.htm

Publications and CAD Communities

What is a community? For some it is a conversation of shared interests. For others it is access to past experience, experts or resources. With a mix of traditional and innovative community gathering, communities like those provided at ConnectPress allow community member's access to like-minded users. All members play a role in the community: casual or peripheral affiliates, leaders, buyers, sellers, promoters, experts, and regular members. Currently ConnectPress hosts communities for almost every major CAD system.

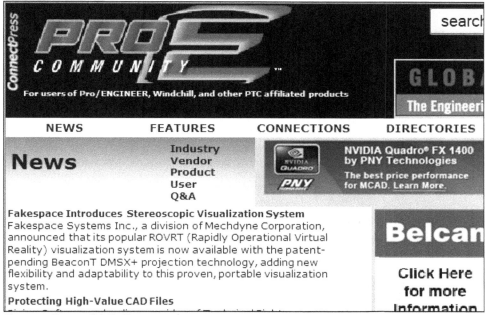

ConnectPress CAD Community www.connectpress.com

Bringing it all together

No single solution works for every situation. A combination of training from: live classrooms, virtual classrooms, web-based, online CBT, VAR's, public/private educational institutions, other resources, and possibly your own company's in-house trainings should be tapped. The number of employees at your facility and their present level of CAD skill should be the starting point for your analysis. Remember, this is not just a one time process. You must be willing and able to initiate a complete training program that will grow and adapt as your needs change. This concept is very similar to that which governs the products which your company designs. Your designs must grow and adapt to the needs of your customers, as should the education/training of your employees in the use of the 3D CAD tools used to achieve those design.

HOME
BACK ISSUES
EVENTS
SEARCH
LINKS
CONTACT US
SUBSCRIPTIONS
RECRUITMENT

ARCH

earch...

Click Here to download this month's MCAD magazine or back sues in pdf format

OST READ ARTICLES

Inventor 11:
Exclusive!
High-end
graphics cards
SolidWorks 2006
– part I
Software: Solid
Edge V17 – part
II
Selecting a
CAD/CAM system
Inventor Series
10: part 1
Inventor Series
10: part II

Now online: November issue
For previous months, visit the back issue section

NX Knowledge Driven Automation

Monday, 06 November 2006

Over the last few releases UGS has been quietly redeveloping its knowledge-driven automation tools within the NX portfolio. Al Dean takes a look at where the state of the art is at.

Anyone that's been following the 3D CAD, PLM, 'call it what you will' industry for the past ten years will know that many vendors and their users are examining the technology they use to find additional ways to optimise their efficiency and productivity...

UGS PLM Europe Event 2006

Monday, 06 November 2006

UGS hosted its annual user event in Frankfurt at the start of October. With the audience nearly double that of last year's, MCAD attended the keynote presentations and exhibition.

Interview with Carl Bass: part 1

Monday, 06 November 2006

In October, Autodesk held its now yearly event for users in Hammersmith, London. Martyn Day caught up with Autodesk's CEO, who had flown in for the event, to talk about CAD file formats, among other industry issues.

Escalating benefits for Kone

Monday, 06 November 2006

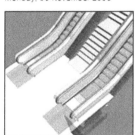

When Kone escalators in Keighley, West Yorkshire, moved to 3D with Autodesk Inventor and MSC.Nastran the design department looked to streamline its many disparate design processes.

MCAD CAD publications online www.mcadcentral.com

Chapter Eight
Selecting 3D CAD software and hardware

- How to select 3D CAD software
- Factors that go into the selection process
- What are my system selection parameters
- Essential aspects of hardware selection

You want the best available 3D product design package on the market, plain and simple. How can you determine – before making the investment – that a 3D CAD solution can meet your team's specific requirements? Prior to selecting and purchasing 3D CAD software, attempt to answer the following questions:

Can our current workforce be productive in a period of time that will adversely affect productivity?
How many workers in your company use CAD in their daily assignments?
Will new hires be required to use, implement, and maintain the new 3D CAD system?
What new hardware requirements will be necessary based on our software selection, including maintenance?
What is the estimated cost of transition including training?
Can legacy data (existing designs) be used to generate 3D CAD models? Are there integrated tools that enable the use of 2D data to generate new 3D models?
Will the software include engineering design and analysis functions such as: mechanism, thermal, stress? Does the 3D CAD package have an integrated product for these functions or do we have to purchase 3rd party software packages to supplement?
What process is used for our enterprise: casting, injection molding, machining? Will the new 3D software accommodate these processes with available modules: i.e. Sheetmetal, surface design, mold, cabling, piping, etc.?
Will we use the 3D model design for downstream processes such as manufacturing, NC, CNC, etc.? What CAM system do we presently use? Do we need to purchase 3rd party software packages for manufacturing? Does the new 3D CAD software come with integrated manufacturing software?
When running benchmarks, do the 3D CAD packages really meet our present needs? Can the software meet future product requirements?

Software selection

Cost is a key factor in upgrading from 2D to 3D CAD software. But the typical license cost of entry-level 3D software can be misleading, since the more important issue is total cost of ownership. Cost of ownership takes into account the cost of software, training and customization, plus the cost and quality of support. So, what is the *total* cost of ownership of the new 3D tool, including the software, associated training, and maintenance?

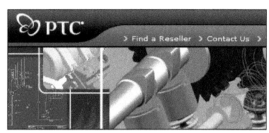

www.ptc.com

While typical mid-level 3D systems are priced at under $5,000 (US), you must be aware of possible hidden costs. The 3D system should be simple to use, so that training costs can be contained, and that users can be productive using the system in as short a time as humanly possible. Also, customizing the tool – that is, building user-specific interface functions and other system preferences – must be easy enough for end users to perform. Doing this with some 3D systems requires third-party assistance, and, of course, incremental cost.

You need CAD tools that are intuitive, yet do not force you to compromise your product, process, or people. Despite outward appearances, there's a vast difference between CAD applications on the market today. Many so-called 'easy' tools are just that, mainly because they're lacking the depth and breadth of functionality you need to get the entire job done.

Your CAD tools need to provide speed, reliability, and innovation. Before you invest in a 3D solution, you need assurance that the vendor you select knows exactly what you need in the trenches, and will continue to invest in cutting edge CAD tools to meet your evolving needs. As you add more projects, more people, and more products in your pipeline, will your CAD provider be a partner who can help you solve tomorrow's challenges?

Some 3D CAD vendors provide support through third parties (VARS). It's best to choose a solution from a vendor that provides his or her own technical support, directly to the end-user. In this case, users spend less time waiting for proper support, and more time using their systems. Since time is money, time saved waiting for answers means money saved in product development.

Since you are considering moving from 2D CAD applications like AutoCAD® to a 3D design tool, you need to know upfront that you can easily leverage your 2D CAD data without forcing engineers to spend days and weeks repairing data. With some 3D CAD tools, 2D data import, conversion, and repair can be a nightmare – but it shouldn't be. Previous chapters in this book have addressed the process of using 2D data in a 3D CAD system.

Choose a 3D tool that has an integrated 2D import wizard that will give you control during the import of AutoCAD and other 2D data. The system you choose should allow the importing of your native 2D data as well as data in the form of: IGES, STEP, DXF, ACIS, etc.

www.solid-edge.com

Is the 3D CAD system you are considering scaleable, can you do simple through advanced design? As your business changes, the 3D CAD system you select should satisfy all your needs for more complex assemblies and PDM, etc. Although entry-level 3D CAD systems are not usually considered appropriate for "high-end" modeling work, over time users may bump up against the limits of their technologies. For instance, a designer may want to work with more complex geometries, or manage larger and larger 3D assemblies.

www.autodesk.com

When this happens, the need for a scalable 3D growth path – becomes clear. Most entry-level 3D CAD systems do not offer such a scalable growth path. To move to the next-higher product class, users are forced to migrate files through a neutral format, such as IGES, to a completely different 3D system. Since these formats cannot carry forward certain types of information, users face severe limitations – and have to spend frustrating hours, days, sometimes weeks repairing their files. The quality of IGES/STEP translation engines varies between 3D CAD systems. Some systems let you define ambiguous/bad geometry that simply cannot be translated. So even though you may get IGES/STEP export capabilities, it may not work properly.

In other cases, entry-level 3D CAD systems may support third-party applications as higher-level add-ons, but such architectures introduce substantial complexity into an already complex endeavor. For instance, the users must now coordinate new-product installs and version upgrades by themselves with a multiplicity of other software vendors.

It's more effective to choose a scalable, integrated solution that allows designers and engineers to spend their time developing products rather than integrating separate software applications. There are systems that provide an affordable entry-level package, and easily expand as your business and technical needs grow, seamlessly integrating with the system you originally invested in.

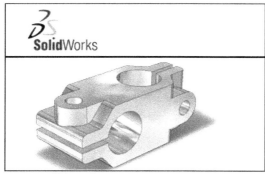

www.solidworks.com

If creating and managing assemblies – small and large – is critical to your job and your business, you cannot afford to compromise your product and your process with a CAD tool that can barely handle 500 to 1000 parts. Yet, that's the limitation of many mid-range CAD tools on the market.

The creation of accurate, detailed drawings from your models so that you can clearly communicate design intent to production is also essential. If your models do not provide visible dimensions and labels, your production team can't create the part as intended – which means more work for everyone. Inaccurate or incomplete drawings can have disastrous results in costly scrap, lower quality products, missed delivery dates, unmanufacturable products, and unhappy customers. When it comes to drawing capability, investigate all aspects of the 3D CAD systems that you are comparing.

To share concepts and ideas with partners and colleagues across your product development teams via the Web, select a 3D CAD system that has built-in collaboration tools. At the same time, you need to protect your company's intellectual property.

Another capability to consider involves the search for existing parts and assemblies, both within and outside your organization via the Intranet/Internet.

www.ugs.com

You can't afford to waste time and resources searching for the CAD tools you need, then getting them to communicate and work together. Yet some CAD vendors force you to search for 3rd-party modules that lack the same technology kernel as other applications you have on board.

At the end of the day, you need a complete set of integrated applications, built on a single platform – not a disparate set of applications that require you to perform complex system upgrades and integrations. You want a single, scalable platform – from a single CAD software package.

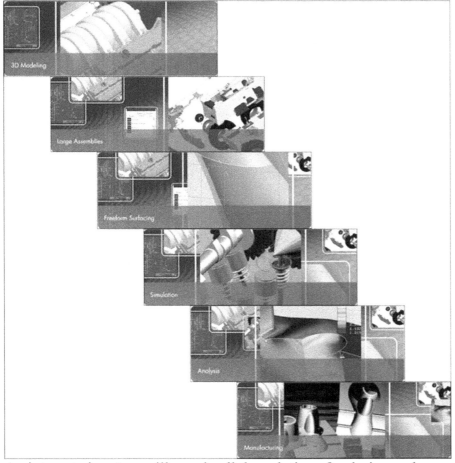

An integrated system will supply all the solutions for design and manufacturing

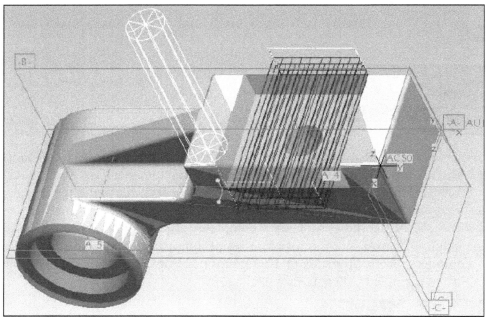

Machining a 3D CAD part with Pro/NC

As an example, most mid-level 3D CAD systems require external manufacturing (CAM) programs. The typical multi-axis CAM program costs $5,000 to $20,000 (US) depending on the number of axis machining capabilities. And though an integrated CAD system will require increased cost for supplemental modules, the costs are small compared to the non-integrated systems that also involve external software integration.

When an ECO is issued for a change to a part designed with an integrated CAM program the change can be made in the part, assembly, or manufacturing model. The change will propagate throughout the design regardless of where it is initiated; part level, assembly level, or manufacturing level.

With a non-integrated system, ECO's require separate databases for the design and the manufacturing models to be reconciled, introducing a multitude of possible errors and costly delays. In many cases the NC sequences and processes need to be completely revamped. Not so with an integrated CAD/CAM system. Changes to a part will propagate throughout the milling sequences as per the geometry changes. Editing, if required, will be limited and downtime kept to a minimum.

The online tutorial for Chapter 3 provides an example of taking a design from part-to-art. The seamless creation of part geometry, detailed documentation, and manufactured part is showcased.

Hardware requirements

When your design team settles on a 3D CAD software package, the next step is to determine the hardware requirements. One of the most typical mistakes made by companies is the selection of hardware without consulting the software company as to their system requirements. There are many in IT (Information Technologies) at both companies and schools that think they know better than anyone else as to the selection of hardware.

A former administrator and co-worker laughed at me when I stated in a meeting a number of years ago that CAD hardware needed to be selected with the consultation of the CAD software company. Their miscue resulted in the purchase of computer stations that lasted only two years. At our school, I was locked out of the selection process only one time over a 25-year period. That lockout resulted in the purchase of monitors that did not support shaded color images. All monitors had to be returned at a cost of $50 (US) per unit restocking fee. The incorrect purchase of computers resulted in having to eliminate PCB and IC courses since the "new" computers did not support these functions. The resulting damage both to our budget and to our reputation for having excellent facilities took years to repair. I want to drive home the possibility of individuals outside the engineering department attempting to determine your hardware. ***Just remember: software drives the selection of hardware.*** I always specify the correct hardware by using these simple three steps:

- Admit that I do not know anything.
- Look for the people who do know (at the CAD software company).
- Get their input and adhere to their recommendations.

IT department should not be allowed to select an engineering workstation. They need to be included, but, they are not to be the sole input in the decision making process. In my experience, IT would select the simplest most familiar hardware regardless of its functionality. Familiarity with the hardware platform means less work for the IT department. IT participation is essential for success, but it should not be the driving force.

Engineering Workstation TriStar PowerStation

Your budget will of course drive many of the selections you make in both software and hardware. Hardware selection means channeling your dollars into the best possible solution for your business, not some CAD software company's idea of what you should have.

Along the same lines, do not let the hardware manufacturer sell you on expensive boxes with enough RAM to design ocean vessels, unless of course you are designing oceans vessels, or airplanes, etc. A common mistake is to sink huge amounts of capital into the box and leave the screen as an afterthought. Your business is designing. That means graphics. That means the larger the screen the easier to design, to visualize, and to display the required 3D CAD graphics. A 21-inch screen should be the minimum. 22-inch and even 24-inch flat panel screens are not that expensive compared to just a few years ago. Also, you want to make your users happy? Want them to sit in that chair for hours on end? Then, get them a 22 to 24-inch (or larger) flat panel and watch their productivity rise before your eyes. There is nothing more seductive to an engineer or designer than seeing their work displayed in 3D on a huge beautiful color flat panel screen. Skimp on the screen and you lose some of the productivity gain you hoped to get from switching to 3D CAD.

Bigger is better Dell Flat Panel

Overall, you need to select hardware as per the minimum requirements stated in the software documentation. Pro/ENGINEER, SolidWorks, Inventor, and SolidEdge minimum requirements are provided here. Since many if not all existing 2D CAD designers use AutoCAD, the suggested minimum requirements for that system are also listed.

In general, if you are running 2D CAD software, your systems may not be sufficient for the switch to 3D CAD. There will be considerable increases in system functionality and demands. The demands on your designer's workstations need to be understood by your transition team.

Some of the factors in your decision to move from 2D CAD to 3D CAD include these considerations:

- The software selected
- OS (Operating System) Type: Windows-based or UNIX-Linux
- Your budget
- Number of new stations required
- Number of existing stations that can be upgraded

The system requirements here were taken directly from each company's website in the fall of 2006. We will start with the system requirements suggested by AutoDesk to run their AutoCAD software:

AutoCAD System Requirements www.autodesk.com

Recommended System Requirements:

AutoCAD 2007 for users who are focusing on 2D drawing creation are as follows:

- Intel® Pentium® IV processor recommended
- Microsoft® Windows® XP Professional or Home Edition (SP1 or SP2), Windows XP Tablet PC Edition (SP2), or Windows 2000 (SP3 or SP4)
- 512 MB RAM
- 750 MB free disk space for installation
- 1024x768 VGA display with true color
- Microsoft® Internet Explorer 6.0 (SP1 or higher)

AutoCAD 2007 for users who are taking advantage of the new conceptual design capabilities are as follows:

- Intel® processor 3.0 GHz or greater
- Windows XP Professional (SP2)
- 2 GB RAM or greater
- 2 GB of disk space available, not including installation
- 1280x1024 32-bit color video display adapter (true color)
- 128 MB or greater, OpenGL®-compatible workstation class graphics card (Get information on graphics hardware certified for use with AutoCAD 2007)

Inventor System Requirements www.autodesk.com

Part & assembly design (less than 1,000 parts):

- Intel® Pentium® 4, Xeon™, or AMD Athlon™, 2GHz or better processor
- 3.5+ GB free hard drive space (for installation)
- 1+ GB RAM
- 128+ MB DirectX or OpenGL capable graphics card

Large assembly design (more than 1,000 parts):

- Intel® Pentium® 4, Xeon™, or AMD Opteron™, 3GHz or better processor
- 3.5+ GB free hard drive space (for installation)
- 3+ GB RAM
- 128+ MB DirectX or OpenGL capable, Workstation Class
- Supported Operating Systems
    - Windows® 2000 Professional SP4
    - Windows® XP Professional SP1, SP2
    - Windows® XP Professional x64 Edition

SolidEdge System Requirements www.solidedge.com

- Windows XP Professional® operating system (32 bit or 64 bit).
- Internet Explorer 6.0 or higher

Hardware System Requirements:

- Intel Xeon or AMD Opteron processor
- Windows XP Professional Service Pack 2
- 1 GB or more RAM
- True Color (32-bit) or 16 million colors (24-bit)
- Screen resolution of 1280x1024

Minimum System Configuration:

- Intel Pentium, Intel Xeon, AMD Athlon, or AMD Opteron
- Windows XP Professional
- 512 MB RAM
- Minimum Resolution: 1024x768
- 65K colors
- Disk space required for installation: 1.3 GB
- The 64-bit version of SolidEdge requires Microsoft Windows XP Professional x64 Edition with Intel EM64T or AMD64 processors

Pro/ENGINEER System Requirements www.ptc.com

Recommended System Requirements:

		Windows (XP and 2000)		UNIX and Linux	
		Minimum	Recommended	Minimum	Recommended
Main Memory		256 MB	1024 MB or higher	256 MB	1024 MB or higher
Available Disk Space	Pro/ENGINEER and Conferencing	2.0 GB	2.5 GB or higher [a]	2.5 GB	3.0 GB or higher
	Pro/ENGINEER and Conferencing with Pro/ENGINEER Mechanica (Structural and Thermal Simulation) Wildfire 3.0	2.0 GB	3.0 GB or higher [a]	3.0 GB or higher	3.5 GB or higher
Swap Space		500 MB	2048 MB or higher	500 MB	2048 MB or higher
CPU speed		500 MHz	2.4 GHz or higher	See above table for individual vendor processor support	
Internal Browser Support		Microsoft Internet Explorer 6.0 SP1 or later		Browser (Mozilla 1.7.3) is embedded in Pro/Engineer on UNIX platform	
External Browser Support		Mozilla 1.7.3 IE 6.0 SP1 and later		Mozilla 1.7.3	
Monitor		1024 x 768 (or higher) resolution support with 24-bit or greater color		1024 x 768 (or higher) resolution support with 24-bit or greater color	
Network		Microsoft TCP/IP Ethernet Network Adapter		TCP/IP Ethernet Network Adapter	
Mouse		Microsoft-approved 3-button mouse		3-button mouse	
File systems		NTFS		All vendor-supported file systems.	
Misc.		CD-ROM or DVD drive		CD ROM or DVD drive	

NOTES
[a] For Windows XP only. For 32-bit operating systems (including Windows 2000), the Windows limit is 2.0GB. For Windows XP you must enable the /3GB switch in order to utilize RAM greater than 2.0GB.

SolidWorks System Requirements www.solidworks.com

Recommended System Requirements:

For Mechanical Design (SolidWorks), Design Validation (COSMOS), and Collaboration (eDrawings) (3).

Microsoft® Windows® Supported Operating Systems (9)

SolidWorks 2004	SolidWorks 2005	SolidWorks 2006	SolidWorks 2007
XP Professional (32-bit) (1)	XP Professional (32-bit) (1)	XP Professional (32-bit) (1)	XP Professional (32-bit) (1)
2000 Professional (2)	2000 Professional (2)	2000 Professional (2)	Not supported
Not supported	XP Professional (64-bit) (4)(5)(7)	XP Professional (64-bit) (4)	XP Professional (64-bit)

Computer and Software Requirements:

RAM   Minimum: 512MB RAM
      Parts (6) (< 200 features) and assemblies (< 1000 components)

      Recommended: 1GB or more
      Parts (6) (> 200 features) and assemblies (> 1000 components)

      Recommended (Very large models): X64 processor with 6GB or more
      Very large models that exceed the 2GB process size limit of 32 bit architecture.
      Parts (6) (> 1000 features) and assemblies (> 10000 components)

      Virtual memory recommended to 2X the amount of RAM

Video   Recommended: A certified OpenGL workstation graphics card and driver
      For a listing of tested and certified graphics cards and driver combinations, visit the Graphics Cards and Systems web site

CPU   Intel® Pentium™ (8), Intel® Xeon™ (8), and Intel® Core™.
      AMD® Athlon™, AMD® Opteron™, and AMD® Turion™

Notes:
(1) Recommended: Windows XP Professional SP2
(2) Recommended: Windows 2000 Professional SP4
(3) eDrawings and eDrawings Professional support the following operating systems; Windows XP Professional 32-bit(1) ,Windows XP Professional 64-bit (1)(5)(6), Windows 2000 Professional (1), Windows Server 2003, Mac OS 10.3.3 Panther and higher, Windows NT 4.0 (Service Pack 6 and higher),Windows 98 SE, and Windows ME.
(4) Supported as a 32-bit application running in the 64-bit OS.
   Supported as a 64-bit application SolidWorks 2006 Sp1.1 or higher.
(5) Supported SolidWorks 2005 SP4 and higher.
(6) Memory usage can be affected by the number of features, complexity of the features, and use of imported data.
(7) COSMOS Designer 2005 can only be installed on 64-bit machines from the SP4 CD.
(8) Support included for Intel® EM64T.
(9) The Windows operating system requirements follow the Microsoft Windows life-cycle policy which can be viewed at: http://support.microsoft.com/gp/lifeselect.

What does my company really need?

3D CAD solid modeling requires more computing power than 2D CAD systems. Purchasing (or leasing) high-quality expandable and upgradeable engineering workstations will ensure a decent length to their work life.

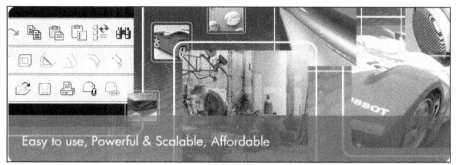

Pro/ENGINEER www.ptc.com

Strictly speaking, a decent quality expandable engineering workstation will run between $2,000 and $3,000 (US) depending on the size of the flat panel screen and the quality of the graphics card.

Since the future is here, Microsoft's Vista operating system or at the least XP 64-bit professional is a must. The same can be said for buying dual chips with dual core processors and 64-bit.

RAM requirements will be dependant on the size of your CAD files. The larger the assembly, the more RAM is required. A minimum of 1 GB RAM is essential, since 3D CAD is memory intensive.

Hard drives have become extremely inexpensive. One 80 GB drive should provide sufficient local storage for your CAD files and other needs.

Graphics Cards

Regardless of how good your processor is, the size of your drives and screen, or the quality of the software, without a 3D CAD level graphics card you cannot expect proper high-quality display of your engineering designs. Poor graphics card selection will negate any increase in RAM, or processor speed.

128 MB RAM is the minimum for the graphics card. In general, if all this is confusing to you, then just look at the suggested minimum card requirements listed for every 3D CAD system and buy a certified card at least as good, and better if your budget will allow.

The most important aspect of graphics card selection is to get a professional "openGL" card and not a gaming "openGL" card. Not all "openGL" implementations are equal. Being a "certified" card is very important. Almost all cards now say openGL, but only those certified by the 3D CAD company (who's software you are purchasing) will work properly.

Pro/ENGINEER www.ptc.com

Memory

Since 3D CAD software is memory intensive, you should plan on a minimum of 1 GB RAM. Using a 64-bit operating system with a 64-bit version of 3D CAD software will also ensure that you are ready for the highest level of performance available. The size of the CAD files your company will be creating should determine the minimum memory requirements. Larger assemblies are very memory intensive.

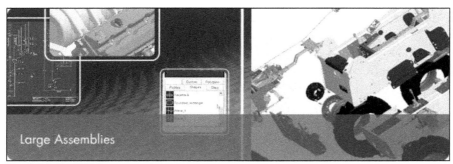

Pro/ENGINEER www.ptc.com

Networking

Your IT people will of course drive the specifications for networking. Productivity can be reduced if the network cannot handle the 3D CAD file system efficiently. Network stability will make or break your engineering department. Network card, router, and switch performance needs to be optimized to have a seamless design environment. The engineer and designer using 3D CAD should not have to deal with a slow or overburdened network.

Dedicated servers for engineering will allow collaboration between engineering departments and sites. Networks without dedicated engineering servers are subject to slow downs and bottlenecks that severely cut into the productivity gains 3D CAD should bring to your companies design staff.

Suggested minimum hardware specifications

Though we hesitate to specify exact hardware requirements for every possible situation, a suggested checklist is provided for your reference. It is preferred that your transition team meets with 3D CAD company representatives or their VARs to discuss any hardware requirements before any decisions are made. Networking requirements are not specified here.

Windows-based

- Microsoft Vista or XP professional operating system
- Two Intel Pentium, Intel Xeon, AMD Athlon, or AMD Opteron 64-bit Dual-Core processors
- 1 GB RAM expandable to 5 GB
- 21 to 24-inch flat panel monitor
- 1600 x 1200 or higher with 24-bit or greater color
- True Color (32-bit) or 16 million colors (24-bit)
- CD/DVD drive
- Certified openGL workstation graphics card and driver with minimum of 128 RAM
- 80 GB HD
- 3-button mouse

Dell Precision Workstation 690 and 490

Dell ratchets up its Precision Workstations a couple of notches by adding dual core Xeon processors to the mix. The leader of the pack is the Dell Precision Workstation 690, which can be stoked with a pair of dual core Intel Xeon 5080 processors at 3.73GHz and your choice of 32-bit or 64-bit Windows XP. Add to that a 512MB NVIDIA Quadro FX 4500 graphics card that can pump out dual monitor DVI goodness, and you're loaded for bear.

Dell's Precision Workstations www.dell.com

UNIX-Linux-based

These are limited to Pro/ENGINEER, Unigraphics, or Catia, since most mid-range CAD systems do not run on UNIX. Here it is advisable to strictly adhere to the 3D CAD companies suggested hardware configuration. Since Pro/ENGINEER is the only "mid range and high end" system, you have the option of Windows or UNIX-Linux.

A high-powered Apple running Windows is also an alternative

Chapter Nine **What you are missing**

- Available down-stream capabilities
- The importance of an integrated CAD package

Originally, I was going to name this chapter: Additional Topics, then I realized that it's more than just additional; it forms the core of 3D CAD design capabilities. In reality, staying with 2D CAD means these are the areas of functionality and specialized modules that you will be missing. When the design process starts and ends in 2D, none of the capabilities described in this chapter are available let alone possible. There is no reason to compare functionality between 2D and 3D since only the 3D CAD database can be utilized for computer-aided manufacturing, free-form surface and mold design, rapid prototyping, analysis, simulation, rendering, or data management, etc.

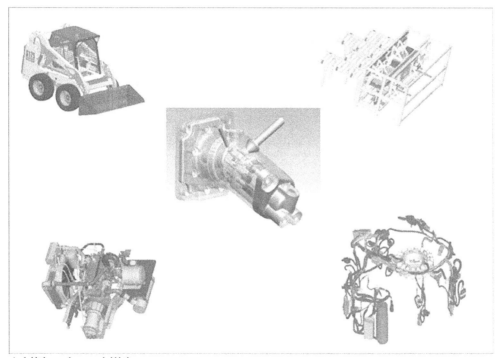

Additional capabilities

As described throughout this book, the 3D part geometry forms the core of the design process. All other applications and downstream capabilities flow from the part's 3D database. Downstream such as prototyping, manufacturing, inspection, and illustration take the part and leverage its database. Concurrent product development means that the part's database can be utilized before the design is complete.

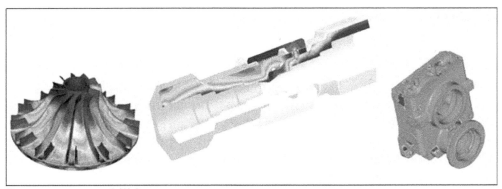

Downstream capabilities

Since few designs go through the design-through-manufacturing sequence without changes, modifications can be initiated at any point in the process. The edits can be propagated vertically and horizontally throughout the design. Changes made in the part will update all other processes. This is the same as changes made in manufacturing can propagate vertically through the part and assembly, etc. The design process becomes more fluid and natural, somewhat like having a "do over" as a kid when playing a game. Here the game is to get the part to market. "do over" turns into a redo with 2D CAD because of the massive amount of work required to document and implement a change. Designs become more rigid in the 2D process because of the fear that changes may actually inhibit hitting the production deadline.

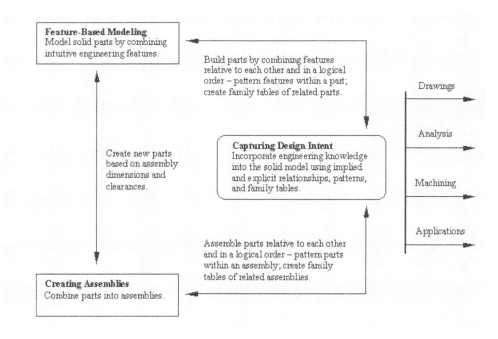

CAM (Computer Aided Manufacturing)

3D CAD systems have harnessed the power of our industry standard NC Machining tools into a single, comprehensive NC solution including support for multi-axis milling machines and multi-axis lathes. With 3D CAD complete machining, manufacturing engineers and machinists now have the most powerful and complete package of NC (Numerical Control) programming capabilities and tool libraries possible.

Surfacing machining

With a high-end integrated 3D CAD system complete machining, manufacturing engineers can work concurrently with designers to automatically incorporate design changes. With this integrated collaboration between two fundamental areas of development, you have the power to increase product quality, reduce scrap, and shave production time and costs.

Complex multi-axis machining

Intense global competition is driving manufacturing engineers to make a greater contribution towards quality, innovation and speed-to-market.

Because of its seamless integration with all 3D CAD tools, an internal CAM system dramatically improves your productivity by eliminating data translation between CAD/CAM applications, automating many programming tasks that can currently take you hours, and leveraging your 3D models to create an optimal tool path.

The key benefits of an integrated CAD/CAM system include:

- Complete integration of design and manufacturing will simplify the creation of manufacturing components (jig and fixtures) and geometry
- Feature-based and geometry based programming, for easy adaptability to design changes
- Predictable and reliable machining accelerates delivery of products to customers
- Capture and reuse your machining practices to streamline and standardize manufacturing methodologies
- A complete solution, from design through NC code generation: NC program creation, process documentation, post-processing, and tool path verification and simulation
- A Fully Integrated CAD/CAM Environment
- Complete associativity across applications
- Parametric and feature-based
- Complete product model definition
- 2-axis prismatic part tool path generation
- Automatic drilling
- Multi-surface 3-axis milling tool path generation

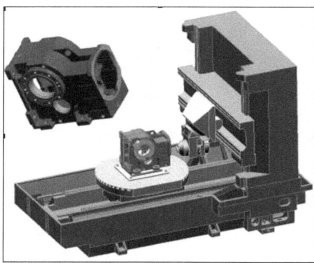

Jig and fixture

Surface/Product Design

By combining the power of parametric modeling with the flexibility of free-form surfacing, you can now create complex, free-form curves and surfaces directly within a single, intuitive, and interactive design environment. Designers and engineers can create conceptual designs and free-form surfaces while having the ability to model the specific engineered components essential in every successful product.

Surface design

With a high-level 3D CAD surface package you can:

- Build free-form geometry at any point in the design, using as many or as few constraints as desired, for maximum design flexibility
- Full associativity allows the surfaces and curves to instantly adapt to design changes
- Create 3D curves by specifying interpolation or control points in one or more views
- Set up references dynamically by snapping to any object
- Create planar curves referencing a plane or radial to another curve
- Create Curve-on-Surface (COS); Sketch-on-Surface; Project-on-Surface
- Move control points dynamically or numerically
- Edit multiple curves simultaneously
- Interactively delete or change references to any object
- Modify tangent constraints dynamically or numerically
- Connect curves and surfaces with positional, tangent, and curvature continuity
- Add interpolation or control points, interactively
- Extend dynamically, with or without constraints
- Delete individual points or curve segments
- Combine and split curves

- View dynamic curve and surface analysis
- Change curve types from free to planar or COS
- Share and manage assembly design data efficiently using predefined skeleton model interfaces
- Define and automate the enforcement of design rules, ensuring that only appropriate relationships are created within the context of the design, so they can be reused easily
- Regenerate surfaces in real time
- Make automatic surface connections
- Reshape surfaces by editing the defining curves
- Add or remove multiple internal curves in two directions
- Replace boundary curves/edges to redefine surface shape
- Change surface types between boundary, loft, and blend while maintaining all references
- Trim surfaces
- Work directly off imported geometry, facets, and sample data
- Drive model changes through parametric modifications
- Benefit from downstream use for additional geometry creation, engineering, simulation, and manufacturing
- Import, generate, and filter raw data
- Import geometry, including curves, surfaces, and faceted data
- Create and modify curves

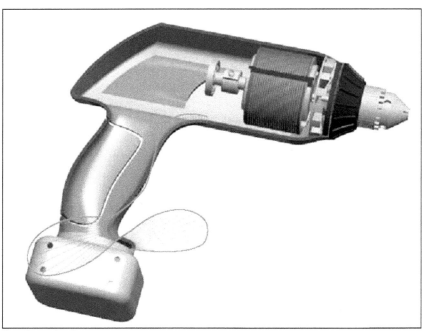

Product design

Tool Design

3D CAD system tool design is the essential 3D CAD tool for professionals who need to rapidly create higher quality mold inserts, casting cavities, and patterns. Using parametric surfacing capabilities, engineers and designers can create even the most complex parting surfaces, regardless of the complexity of their design geometry. By automating many time-consuming, complex processes, 3D CAD systems enable you to focus less on tedious tasks and more on creating innovative, top quality tool designs.

Since the 3D models you create using 3D CAD systems automatically reference your mold and casting designs, any changes you make are instantly reflected in tooling and patterns, which accelerates the product development process.

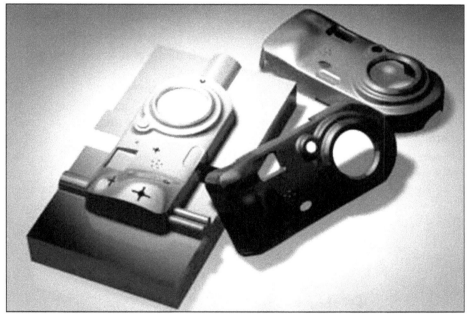

Tooling for molds

High-end 3D CAD can do the following:

- Graphically evaluate mold draft, undercut, thickness and projected area, and then make instant repairs
- Design within two process-driven user interfaces – one for mold, one for casting – each guiding you step-by-step through the process of creating mold and casting cavity and patterns
- Create and modify any features such as drafts, rounds, complex surfaces and parting lines to improve moldability
- Compensate for both isotropic and anisotropic shrinkage
- Build patterns and sand cores that reference design part geometry
- Create parting lines by simply selecting the mold opening direction

- Design parting surfaces, including steel-to-steel shutoff surfaces
- Check for mold lock condition with mold opening and interference checks
- Calculate fill volume
- Split, using the parting surface, and create solid model mold components such as cores, cavities and sliders
- Create multi-cavity layout configurations, including single, rectangular, circular, and variable
- Automate the placement, trimming, and clearance of holes for over 9,000 different ejector pins
- Select and quickly assemble user-customized injection molding machine mock-ups in order to check for possible interference
- Create waterlines and instantly analyze for thin wall conditions
- Simulate the mold opening sequence, including interference-checking
- Generate production-quality detail drawings, including BOMs and balloon notes
- Produce runners, gates, and sprues instantly
- Automates creating parting surfaces
- Compensates for model shrinkage by enabling you to dimension or scale the entire model in X, Y and Z
- Seamlessly integrates with 3D CAD system for mold simulation
- Produces solid models of inserts that maintain an associative link to NC applications; if the design part changes, the mold inserts and NC tool paths automatically update
- Eliminates the need to translate between part design, mold design, and NC, due to seamless integration with other 3D CAD system applications
- Erases costly rework from interference-checking and mold-opening simulation
- Enables new users to become productive immediately, with easy-to-use interfaces both for mold and casting

Tool design

Mold Design

One of the primary challenges facing manufacturing engineers today is finding time to improve quality, speed and innovation in mold base design and detailing. Mold Design lets you create extract components that you can then use to create mold details during actual mold production. After creating a mold model, you can use these extracted mold components with NC Manufacturing to create CNC tool paths.

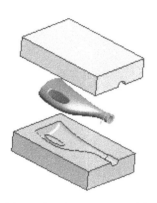 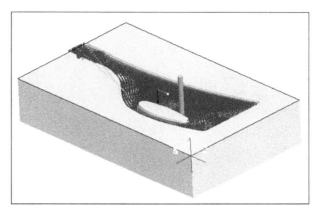

Mold production and machining

You can create final extract components that reflect the geometry of the design model along with shrinkage considerations, adequate drafting, ejector pinholes, runners, and cooling systems. You can also verify that the components do not interfere with each other during the mold opening process. You can also perform different tests to check validation of your design: draft check, thickness check, projection area check, waterline clearance check, and components interference check.

Once your components are placed, operations such as clearance cuts and drilled and tapped holes can be automatically created in the appropriate neighboring plates and components, thus eliminating tedious and repetitive mold detailing tasks. It also enables mold-making companies to capture their own unique design standards and best practices directly within the mold assemblies and components. Mold opening simulation, complete with slider, lifter and ejector simulation can be created automatically. Interferences can be checked automatically during the mold opening sequence.

During the molding process, changes to the design model may occur. When these changes are made to the design model, they will propagate throughout all aspects of the design to engineering drawings, finite element models, assembly models, and molding information. Because the mold design engineer references the parametric design model directly, changes are reflected throughout all the intermediate process steps and captured in the molding model.

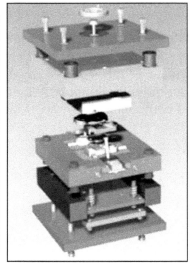

Mold design

In general, a high-end 3D CAD system will enable:

Design part creation
- Create models including features requiring surfacing
- Import and repair geometry if necessary
- Analyze if a design part is moldable
- Generate parting lines and detect undercuts
- Fix problem areas by creating draft, rounds, and other features as needed

Cavity Creation
- Assemble and orient design model while checking draft and projected area
- Apply a shrinkage that corresponds to design part material, geometry, and molding conditions
- Generate the workpiece stock from which core, cavity, and inserts will be split
- Create parting geometry, including sliders, inserts, automatic parting lines, etc.
- Automatically split the workpiece to create cores, cavity, and inserts as solid models

Mold Layout Creation
- Create mold assembly
- Placement and patterning of mold cavities to allow multi-cavity molding
- Online selection and automatic assembly of standard mold bases
- Modification of mold base plates to allow for assembly of mold cavity
- Online selection and automatic assembly of ejector pins and other mold catalog items
- Automated creation of runners
- Automated creation of waterlines, including 3-D waterline interference checks
- Define and simulate mold opening and check for interference between mold components

Drawing Creation
- Create complete production drawings, including dimensions, tolerances, automatic bill of materials (BOMs) with or without balloon notes
- Use of drawing templates

Reverse Engineering

With reverse engineering, you can create – or recreate the electronic 3D CAD model of a physical product. Reverse engineering allows engineers to work with 'point cloud' data obtained by scanning physical prototypes. You can refine the point cloud and polygonal data, reducing noise and/or the total number of points. From there, you can create a facet model, the next step in the re-engineering process. Next, surfaces can be projected to fit facet data or boundary surfaces created from curves sketched on the facet model, or take advantage of the geometric surface options.

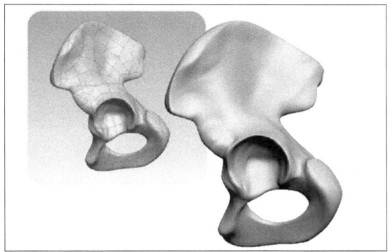

Imported hip scan reversed engineered into 3D CAD model

Once a surface is created, you can use analysis tools for surface analysis, and to check for any deviation between the surface and point cloud. Reverse engineering captures physical characteristics such as surface tangency, ensuring that the design intent is preserved. You can recreate the CAD data for a product for mass-production, or re-use the information for highly customizable products.

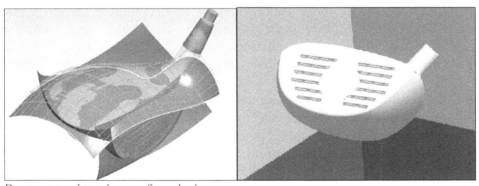

Reverse engineering surface design

Cabling

The process for designing electrical harnesses can be very complex – and problematic. Typically, engineers must create a physical prototype; then manually determine a route for the cabling, which is usually the first route that works, and not necessarily the optimal one. When an engineering change is made to the design, it means starting over by recreating the physical prototype and rerouting the cabling. Additionally, the companies that create the prototypes often do not document the cabling routes, making it difficult to service these products. To solve these issues, a company will often send a service engineer to the field, which can be very costly.

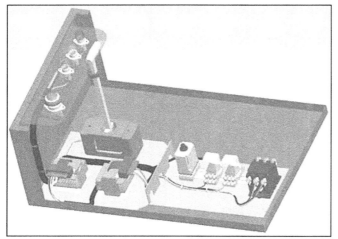

Cabling

A cabling module eliminates these issues by giving you the ability to extract logical information from 2D schematics, while automating your 3D cable routing. You determine a working route quickly, but you can easily find the optimum route that won't interfere with the design.

This combination of tools enables the complete digital model to be defined, thus reducing the dependence on physical prototypes, and therefore significantly lowering product costs. Indirect costs can be reduced as well. Since diagramming tools are included. As part of the design process, mechanical designers will route cables within their 3D assembly using schematic designs as maps. 3D CAD routing capabilities automate this step, extracting the schematic information and electronically driving the routed systems within the 3D MCAD system. This compatibility not only speeds the 3D design significantly by removing the tedious, manual process of interpreting 2D schematic diagrams, but it also eliminates errors by ensuring adherence to the logic defined in the schematics.

Piping

Product complexity is growing rapidly today due to customers' increasing demands and diverse product requirements. This means designers are now challenged with adding more complex piping schematics to a variety of new product designs. Technology has advanced such that engineers now have numerous options in material selection and piping design layout.

3D CAD piping supports all types of industries and styles of piping – streamlining the entire design process, whether you are designing products with hydraulic or pneumatic hoses, high and low pressure tubing, copper work, or even large bore pipes,

Determining the routes of pipes can be a difficult, time-consuming task. Beginning with the creation of a physical prototype and then manually routing the pipes through it, this long, tedious process is typically error-prone. Even worse, it requires building frequent physical prototypes. With 3D CAD piping packages, these problems can be reduced if not avoided altogether.

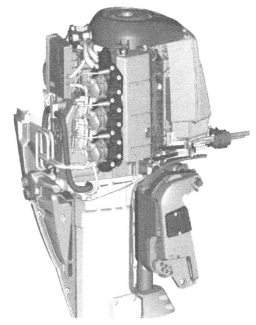

The pipe routing process is now automated. Designers no longer have to build physical prototypes and struggle through the trial and error process; they can determine, virtually, the optimum path of the pipes in the model. In addition, the designer can create rules based on company best practices or corporate policies. The software will then verify that this optimum path is compliant with established rules. Plus, designers have access to a library of standard fittings, which can be reused from product to product, reducing time-to-market and increasing designer productivity.

The fully associative nature of an integrated 3D CAD package ensures that the pipe routing – and accompanying documentation – will automatically update with any design changes made to the model.

Rendering

Rendering packages allow you to produce photorealistic 3D product images for use in design reviews, marketing collateral, technical documentation, user manuals, and product packaging – without having to build a prototype. As your design concept changes and evolves, you can easily update your images without creating a new prototype. A picture is worth a thousand words. That's why companies invest significant time and money into building a physical prototype that can be photographed for use in marketing materials or consumer testing. Some systems have an integrated rendering capability; others require the purchase of an expensive external package. Make sure you get the capability demonstrated using the 3D CAD system functionality, not just a promise that it can be done.

Rendering

     With 3D rendering, you can create images with amazing realism. You can select a specific material, and apply photo-quality properties like a smooth, glossy finish, or a rough, matte finish. You can also represent the product's intended environment realistically.

     With the ability to directly manipulate the lighting through 360° of rotation, you have the flexibility to create the unique effects. You can also create a suitable environment for the product to be rendered in, for example, an outdoor scene or an office desktop. With some systems, you can apply special effects like fog, light scatter, lens flare and depth of field to enhance the final image. These advanced capabilities not only create images that impress consumers, but they also give you back time to optimize your design, and more budget to produce collateral that will generate both sales and excitement around your product.

The image of a showerhead demonstrates the power of rendering, with portions of the part self-reflected within the image; soft shadows realistically represented on the floor; and procedural plastic texturing applied to the knob.

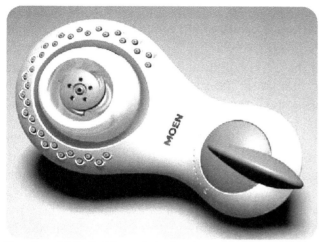

Rendered 3D CAD model of a showerhead

With photorealistic images of the product in its actual environment, design reviews are much faster and generate more enthusiasm. 3D-rendered images can also be used for consumer testing, and for tailoring the product and product launch to ensure success, at a significantly lower cost. High-quality images can also be used in technical publications, such as product documentation, white papers and user manuals.

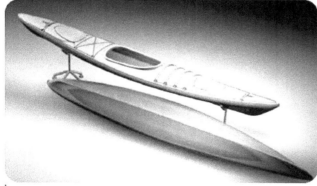

Rendered Kayak

With 3D rendering, your product will break into the marketplace not only with accurate promotional material, but with visuals showcasing all aspects of the product. One key benefit involves improving communication via better product presentation and collateral for design, marketing, training, and customer-facing materials. Increase time for creativity, decrease time waiting for rendering.

Analysis

Today's competitive marketplace is forcing design teams to 'get it right the first time'; the earlier in the development cycle that designers can understand product performance, the faster a quality product gets to market. When teams must rely on costly, time-consuming physical prototyping to test product behavior, schedules and budgets are quickly compromised. True, CAE tools offer a solution, but they're usually disconnected from the CAD solution. Thus, engineers must spend valuable time translating data and preparing the model for analysis. Then, each time there's a design change, designers have to repeat the translation process. Moreover, CAE tools require users to have an extensive specialized skill set.

With an integrated analysis capability, design engineers can better understand product performance and then optimize the digital design – early on in the design cycle, without needing a background in simulation. Integrated systems have the same user interface, workflow and productivity tools that are prevalent throughout the base 3D CAD system. Therefore, product designers can have the same performance and associativity without needing to learn a new program. In addition, integrated analysis systems can analyze native system models and store the analyses in the model files. This means no data translation and data management is more efficient.

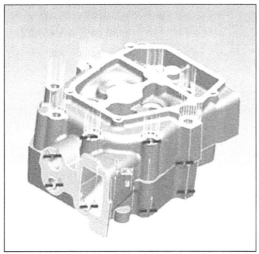

Simulation   Turbine blade analysis

As an example, with 3D CAD analysis tools, you can determine where the higher stress areas are on this turbine and make adjustments to the model as needed.

With the ability to evaluate product performance on-screen, 3D CAD systems give an engineer the freedom to explore new ideas and design variants, and then optimize their designs. Meanwhile, they will have confidence that new designs will satisfy performance requirements, require fewer changes during physical prototyping and deliver superior value.

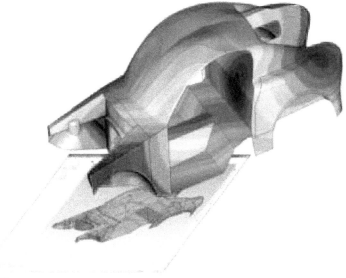

Product evaluation

Key Benefits of an integrated analysis program:

- Gain early insight into product performance and discover design flaws early as you increase first-time build success
- Improve user efficiency with an intuitive, familiar user interface
- Obtain realistic performance data and improve product quality by directly applying real-world conditions to design geometry
- Evaluate more scenarios than with physical prototypes
- Save time and reduce errors by working in a seamlessly integrated design and simulation environment – with no data translation
- Increase innovation by simultaneously designing and simulating design variations
- Decrease development costs by reducing or possibly eliminating physical prototyping
- Provided customizable the simulation processes

Managing Data

Managing a complex product development environment is no longer a problem unique to large manufacturing companies. Small and medium-sized organizations now face these same challenges: increased complexity of products, greater amounts of digital product information, and geographically dispersed teams. The solution to these challenges has historically been to deploy an on-premise, enterprise product lifecycle management (PLM) solution – an option only large companies could pursue. While being able to meet the specific needs of large companies, these enterprise solutions typically require a larger upfront investment, dedicated IT resources and longer deployment timelines.

Product data management

For small and medium-sized companies that recognize the need for PLM, yet either lack the resources – human or financial – or are not interested in the overhead of managing a PLM solution in-house, a PLM package such as Windchill which is integrated into a 3D CAD system is available. It has all the benefits of PLM, such as data vaulting, document management, direct multi-CAD integrations, change control, collaboration, project management and visualization, without all the hurdles. You can improve collaborative project management with customers, partners, suppliers, and outsourced manufacturing without cluttering your PDM solution or giving these groups access to your confidential product information.

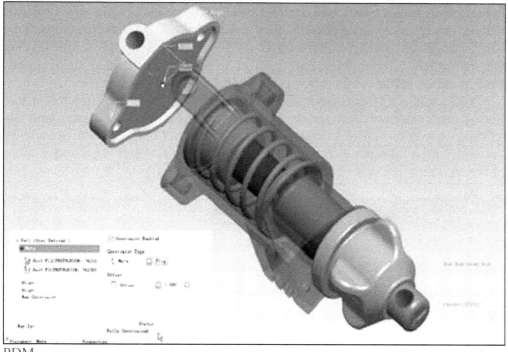

PDM

3D CAD data management solutions can help manufacturers speed development cycles to meet deadlines, respond faster to changing requirements, improve communication to keep everyone on the same page, and lower costs. The solution is tailored specifically to meet the needs of small and medium manufacturers with affordable pricing, easy adoption, and immediate return on investment – without requiring IT resources to deploy and manage the on demand application.

191

Product Data Management (PDM) is a category of computer software that aims to create an automatic link between product data and a database. The information being stored and managed (on one or more file servers) will include engineering data such as CAD models, drawings and their associated documents. "Associated documents" is a collector term for: requirements, specifications, manufacturing plans, assembly plans, test plans and test procedures. The package may also include product visualization data.

The central database will manage metadata such as owner of a file and release status of the components. The package will: control check-in and check-out of the product data to multi-user; carryout engineering change management and release control on all versions/issues of components in a product; build and manipulate the product structure BOM (Bill of Materials) for assemblies; and assist in configurations management of product variants. This enables automatic reports on product costs, etc. Furthermore, PDM enables companies producing complex products to spread product data in to the entire PLM launch-process. This significantly enhances the effectiveness of the launch process.

Product design firms and manufacturing companies are under constant pressure from customers demanding more for less. Too many projects and not enough time are resulting in mistakes and shortcuts that impact costs, schedules, and quality. Last minute changes in customer requirements are making the schedule pressures even more severe.

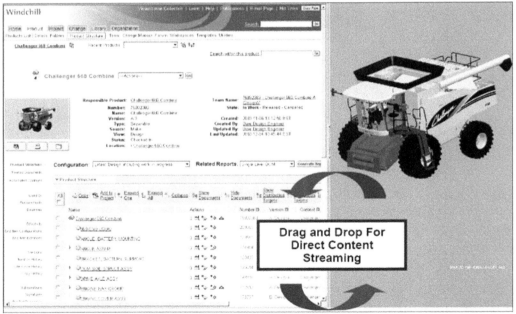

Product lifecycle management

An integrated 3D CAD PDM delivers measurable benefits such as:

- Reduced product development cycle times
- Increased part reuse
- Reduced time spent searching for information
- Reduced engineering rework and scrap
- Reduced engineering change processing time

Product lifecycle management (PLM) is the process of managing the entire lifecycle of a product from its conception, through design and manufacture, to service and disposal. PLM is a set of capabilities that enable an enterprise to effectively and efficiently innovate and manage its products and related services throughout the entire business lifecycle. It is one of the four cornerstones of a corporation's IT digital structure.

Product control

All companies need to manage communications and information with its customers CRM (Customer Relationship Management) and its suppliers (Supply Chain Management) and the resources within the enterprise (Enterprise Resource Planning). In addition, manufacturing engineering companies must also develop, describe, manage and communicate information about their products (PLM).

Documented benefits include:

- Reduced time to market
- Improved product quality
- Reduced prototyping costs
- Savings through the re-use of original data
- A framework for product optimization
- Reduced waste
- Savings through the complete integration of engineering workflows

The product lifecycle goes though many phases and involves many professional disciplines and requires many skills, tools and processes. Product Lifecycle Management (PLM) is more to do with managing descriptions and properties of a product through its development and useful life, mainly from a business/engineering point of view; whereas Product life cycle management (PLC) is to do with the life of a product in the market with respect to business/commercial costs and sales measures.

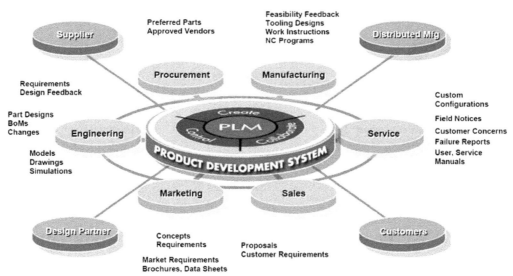

Optimized product development

194

Chapter Ten Success stories

- Custom engine parts company on the road to success
- Free form surfacing embolden bottle design
- Lawn equipment manufacturer improves efficiency

Entrepreneur revs up quickly on 3D CAD
Start-Up Custom Parts Business Turbo-Charges Race Cars
DV/DT Fabrications, Chicago, IL

DV/DT Fabrications is a small fabrication facility operating out of the northwest suburbs of Chicago. The company is a part-time venture of Albert Raczynski, an engineering student at the University of Illinois at Chicago. By day, Raczynski attends engineering classes; by night and on weekends he designs and builds custom engine parts.

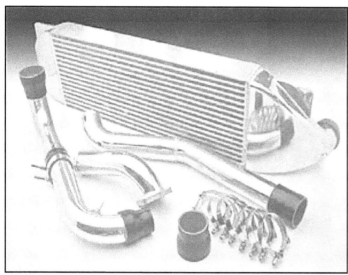

Designing custom parts with 3D CAD

The Challenge: Parlay a Passion for Cars into a Small Business

Raczynski loves cars – loves racing them, designing them, working on them. So when he envisioned a business model that combined his passion for cars with his mechanical engineering aptitude, it seemed like a perfect fit. To bring this business model to life, Raczynski started learning about model design software, then began designing and building custom auto parts in his garage.

The Solution: ramp-up fast on 3D CAD. After taking Pro/ENGINEER Wildfire, PTC's 3D product design software, for a one-day test drive, he discovered the 3D design tools he needed to bring his ideas to life. Using the Integrated Surface Design Extension tool – ISDX, the enterprising new CEO was able to create the complex curves and surfaces he needed to quickly find the perfect look and design for each part.

The Result: On the Road to Success

While many of his peers will be pounding the pavement looking for jobs after graduation, Raczynski will be looking to expand his own fledgling business. Raczynski's newest, most complex product is an aftermarket intercooler kit used for drag racing that boosts performance by optimizing airflow through the engine.

Raczynski is now building a solid reputation via word-of-mouth, through appearances at car races, and through placement of magazine ads, all of which will increase his customer base and enable him to pursue his dream full-time. "With Pro/ENGINEER® Wildfire™, I created the first half of the end tank design in just two hours, and then simply mirrored the part to create the second half. As I started my business, designing the product was the easiest part of the whole process for me." –Albert Raczynski, Founder, DV/DT Fabrications

Custom parts

Venturing into the World of Design

Like many ambitious college students, Albert Raczynski is on the road to achieving a dream. A fourth-year mechanical engineering student at the University of Illinois at Chicago, Raczynski was first exposed to product design at school. In fact, he had recently taken a class where he learned to use AutoCAD.

However, when Raczynski decided to design real auto parts and start a small business, he realized he needed 3D CAD (computer-aided design) skills, the tools of design professionals. Not sure which direction to take, he turned to a tried-and-true source of information: his father. Raczynski's dad, who runs a product design company, relies on Pro/ENGINEER Wildfire for his own work, and encouraged his son to explore the software.

After just one day of 'playing' with the tools, the young Raczynski was convinced that Pro/ENGINEER – along with its surface and styling tool, ISDX – was far easier to use than AutoCAD, with a greater breath of functionality. Hence, he selected Pro/ENGINEER as the design solution for his company, DV/DT Fabrications. "This (intercooler kit) design couldn't have been done in AutoCAD," explains Raczynski. "AutoCAD is good for 2D, but that's about it." The 3D design capabilities in Pro/ENGINEER, along with its flexibility and model preview functionality, made the design process simple. In addition, Pro/ENGINEER's advanced surfacing capabilities made it easy to create the complex curves and surfaces that were critical to the design.

Creating a Breakthrough New Auto Part

Raczynski quickly put Pro/ENGINEER to the test on his hottest new product idea, the 2g/Evo 8 Intercooler Kit designed for the Mitsubishi Eclipse, Lancer Evolution 8, and Eagle Talon. These kits, and other products created by DV/DT Fabrications, are specialized aftermarket auto parts designed to turbo-charge factory-standard engines for drag racing. The intercooler kit is a custom-cast intercooler end tank poured from an aluminum alloy. Because it is cast in aluminum as a single unit, and not fabricated with separate pieces of sheet metal, it is more durable, with considerably better flow characteristics than a fabricated tank. When installed, the intercooler kit boosts a cars performance by reducing the charge air intake temperature in the engine.

The Design Process

Raczynski's idea for this design was derived from observing that most sheet metal end tanks sacrifice intercooler flow in favor of ease of fabrication, universal fit, and mass production. Raczynski wanted to optimize the design of the end tank for evenly distributed flow, a superior fit, and extremely high boost pressures. By custom producing cast aluminum end tanks, he could create a design that was less "boxy" and minimized the loss coefficient associated with bends in the intercooler plumbing system. As a result, the engine and turbocharger could be used more efficiently. The 3D capabilities of Pro/ENGINEER Wildfire made it easy for Raczynski to visualize and manipulate the design of the end tank. The software's flexibility enabled him to quickly experiment with multiple iterations and quickly arrive at the design that would provide optimal air flow and boost performance.

Because the design called for complex surface creation, Raczynski also utilized Pro/ENGINEER Interactive Surface Design Extension (ISDX), a popular tool for creating highly stylized surfaces.

With ISDX, Raczynski was able to effortlessly manipulate the spline curves to optimize the aerodynamic qualities of the surface, and ensure a perfect fit within the tight confines of the car's frame. Having a surface and styling tool that's seamlessly integrated within a 3D CAD design software meant that Raczynski could make any number of changes and instantly see the effects, which added speed and ease to the design process.

"With Pro/ENGINEER Wildfire, I created the first half of the end tank design in two hours, and then simply mirrored the part to create the second half in only 30 minutes," said Raczynski. Remarkably, with the ease of use and intuitive interface of Pro/ENGINEER, he was able to perform this high level of design without any formal training on the 3D CAD system.

Down the Road

After finding the right CAD tools and creating the perfect design, the next challenge for Raczynski was to find a foundry that would work with a start-up company. This proved to be a difficult challenge, but once he discovered the right foundry, Raczynski was able to produce his final design in quantities that were cost effective.

In a detailed experiment conducted in Raczynski's garage with a furnace blower, he tested his newly designed end tank against a typical sheet metal fabricated end tank, and calculated the difference in airflow: The DV/DT design achieved a 23% increase in air flow!

In true collegiate fashion, Raczynski summarizes the results of his testing in a report: "In conclusion, it is the experimenter's belief that a cast smooth volute end tank is superior in performance to any fabricated end tank for flow, fitment, and structural reasons."

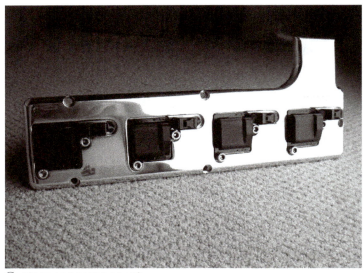

Custom car component

With his deep knowledge of cars, a talent for design, and an entrepreneurial spirit, Raczynski may soon be on the lookout for more real estate, so that he can expand DV/DT Fabrications from a single cylinder, garage-based operation into a successful new small business.

No doubt, the University of Illinois at Chicago and the elder Raczynski are very impressed.

Intake component

O-I uses 3D CAD to design glass containers for the world's top brands
Speeds Process, Improves Quality, and Thrills Customers with 3D Designs
O-I, Harlow Essex, England

In 1903, Michael J. Owens invented the first automatic bottle-making machine, changing forever the way glass containers are manufactured. Today, nearly every glass container is made with O-I's proprietary technologies. O-I glass containers don't just show off the product – they often define the brand of some of the world's most powerful companies, such as Dewar's Scotch, Beefeater Gin, and Lowenbrau Beer. With 100 manufacturing plants on five continents, O-I is known for low-cost, high-quality packaging, enabling customers to showcase their products.

The Challenge: Upgrade from Outdated 2D to 3D CAD

O-I's UK Design Center was designing with 2D software. However, since more and more moldmakers are using 3D CAD, management knew it had to modernize. And because of the limitations of 2D drafting, the engineering team knew there would always be misinterpretations when creating the design.

Added to that, there was a quality issue with 2D. Since the moldmaker had to recreate each 2D design, a very time-consuming process, errors often occurred due to misinterpretation, causing the moldmakers and the design team to go back and forth…time and again. Worse still, with 2D design, additional prototyping was often required, which delayed the finished design by as much as three weeks. With 3D CAD software, the O-I design team hoped to improve the quality of the initial prototype and minimize repeat design efforts.

Bottle design

The Solution

When O-I's headquarters (Toledo, Ohio) made the decision to update its systems globally to Pro/ENGINEER – the 3D CAD design solution from PTC – it allowed the United Kingdom division to make great strides with its design and manufacturing. Before Pro/ENGINEER, it was the moldmakers' responsibility to create a 3D mold from the designer's 2D model. Pro/ENGINEER changed this process completely, and now the original design is created in 3D without any translation, minimizing the chance of interpretation errors. The design team has 'taken back the cavity' of the containers–the trickiest part of making glass bottles.

Working directly with brand owners and their design agencies, O-I designers can translate concepts into reality with exceptional speed and creativity. The design team can now show customers a 3D rotating model on the screen. And, the customers know that "what you see is what you get." This early visualization lets O-I's customers get their new products to market more quickly, creating a competitive advantage.

Brand name bottle design

"With Pro/ENGINEER, I can get a mold cavity done in 15 to 20 minutes versus a couple of days' work with our previous 2D CAD system. Mass properties, associativity, and the analytical capabilities of Pro/ENGINEER have all helped to reduce design time." –Bob Pedder, Mold Design Manager, O-I

O-I's business is built on the knowledge that people prefer glass; everything looks better, feels better, and tastes better in glass. To consumers, glass also connotes premium quality. And with the global demand for glass containers continuing to expand, O-I made the switch to Pro/ENGINEER for worldwide use by its design teams in order to better compete, grow, and profit in the glass container market.

With Pro/ENGINEER, production lead times are continuously falling. In fact, O-I's new product development team in the UK regularly delivers new products to market within 12 weeks of receiving the design brief. The company's use of 3D CAD has been a major factor in offering ultra-short lead times. Because of 3D CAD, O-I has been able to create the UK's first glass design center, which works exclusively in 3D, offering customers greater flexibility, increased accuracy, and faster development times.

Built-in Automation

Keith Reynolds, manager of the design and packaging department, explains that O-I wanted to choose a 3D design tool that would enable knowledge-based engineering. As an advisor on the team that evaluated the engineering tools for all of O-I, Reynolds recalls, "From a design perspective, we needed the ability for models to be parametrically related and relationship-driven with automation on the front end. The development team in the United States conducted a pilot, and we were impressed Pro/ENGINEER built in so much automation."

Another key reason O-I chose to use Pro/ENGINEER on a global basis was because it could let geographically distributed teams share design information. O-I uses the inheritance features of Pro/ENGINEER to share key geometry features with other engineers. For example, engineers in different locations can be working on different parts, and when changes made to one part affect the other engineer's part, that part automatically updates with the modifications. The company is now considering another PTC product, Windchill ProjectLink, to further facilitate extended global collaboration.

Demanding Project Becomes Much Easier

When a London design agency came to O-I with a very demanding project for Dewar's Scotch whisky, it gave O-I's UK team a chance to demonstrate the power of Pro/ENGINEER.

David Sylvester, lead designer, explained that O-I needed to design 13 new bottles while meeting a very tight deadline. Within the design, there were quite a few new elements, including a new, complicated closure that had never been built before. "We knew this would require major changes to the manufacturing process, but with Pro/ENGINEER and our engineering expertise, it didn't turn out to be as difficult as we thought it would," said Sylvester.

The surfacing capabilities in the Pro/ENGINEER Foundation Advantage Package allow O-I to create complex curvatures required in bottle designs. O-I also found the Pro/ENGINEER surface geometry to be so robust that design engineers can easily tweak the design as they iterate to optimize it, or to create new design variants.

The project also called for designing bottles with a variety of capacities and different closures for the different global markets. "Previously, each bottle would have been designed from scratch," said Sylvester. "With Pro/ENGINEER, once we had the master model, we could very quickly make bigger and smaller versions, and assure that the design requirements that the customer had specified would be exactly what they would get in glass."

Designing in Minutes – Instead of Days

With Pro/ENGINEER, the O-I UK design team has become more productive, and is able to deliver finished projects to its customers more quickly. Reynolds states, "We are saving a lot of time downstream in the process now that we are working in 3D. The second, third and fourth variations of a bottle take only 10 minutes to design with Pro/ENGINEER." Previously, each design variation may have taken an extra day.

"Plus, subtle details, like adding the locating marks for where the label is placed, are much easier," he adds. The surface analysis capabilities available in Pro/ENGINEER Foundation Advantage also allowed O-I to analyze how the label would adhere to the bottle. Engineers were able to complete the analysis and tweak the design virtually – without testing it on a physical prototype – which saved both time and costs.

Eliminating Misinterpretation

The quality of the finished product that O-I is delivering is also better. Reynolds explains, "When you are modeling in three dimensions, what you see is what you get. The buck stops here – with my design team." Because the integrity of the 3D design is maintained when molds are created, quality problems due to misinterpretation between the design team and the moldmakers have been virtually eliminated. Finally, because O-I can now share a 3D design with the customer, the team is communicating design intent much more clearly and quickly.

"Everyone has a fantastic reaction to the 3D model. It's a definite 'wow' when they see their product whizzing around on the screen in 3D," he continues. "Now, we have the confidence that this is really what the customer wants." Pro/ENGINEER has delivered significant time-savings and higher levels of quality for O-I. Since models are now parametrically related and relationship-driven, mold cavities can be created directly from the 3D model without the need for translation, and without the dangers of misinterpretation. O-I's designers and mold makers are finding their jobs a bit easier these days, and their customers are thrilled with the results.

Hayter continues its tradition of excellence in 3D design
Premium Lawnmower Manufacturer
Hayter Limited, Spellbrook, England

Improved Design Efficiency with 3D CAD

Founded 60 years ago, Hayter was an early pioneer of the rotary lawn-mower in the United Kingdom, and has since become a household name with both the home gardener and commercial landscaper.

It all started when local builder Douglas Hayter needed to get to the drying sheds where he used to cure wood, and it was often difficult to make his way through the overgrown grass. He solved the problem by mounting a two-stroke motorcycle engine to the top of a dustbin lid, then added some wheels and attached a sharpened blade to the crankshaft – a rotary mower was born. Requests for copies of his rotary mower converted Hayter from homebuilder to mower manufacturer.

Hayter Lawnmower

Today, Hayter mowers are the premium brand in the United Kingdom. The company sells through specialist dealers, and its products are designed for commercial landscape gardeners and contractors, large municipalities, as well as high-end consumer use.

The Challenge: Adopting 3D CAD

In 2003, Hayter's sister company, Murray, decided to upgrade to a 3D design platform. After extensive benchmark testing with at least a dozen vendors, Murray chose the PTC® 3D product design software solution, Pro/ENGINEER® Wildfire™. This design platform was selected based on its functionality and ease of use. Hayter had been considering a change to 3D design, and this corporate decision suddenly made it a reality.

The Solution: Easy-to-Learn 3D CAD

Since Hayter was initiating designs for two important new product lines, there was a real concern about the learning curve for the new software. And since Hayter's engineers were only familiar with 2D design software, they realized they would need to come up to speed quickly on the 3D environment. Hayter engineers are now able to define complex 3D product assemblies.

The Result: Increased Design Productivity

Hayter has achieved a two-month reduction in the design cycle of the mowers' aluminum decks by replacing its 2D design application with Pro/ENGINEER Wildfire. Some of the improvements seen to date include easier visualization of the design in 3D, increased automation with assembly management, and improved aesthetics achieved with Pro/ENGINEER Interactive Surface Design Extension (ISDX). As Hayter engineers continue to utilize the many features of 3D CAD, they anticipate additional efficiencies in the future.

On-the-Job Learning

The seasonal sales cycle for lawnmowers means that it is critical for Hayter to get new products to market on schedule. Therefore, decreasing time-to-market is a constant goal for the engineering team.

With two new product lines in development – the Harrier 56 consumer mower and the MT313 commercial ride-on mower – the pressure was on. When Hayter engineers learned that they would be transitioning from 2D to 3D and learning Pro/ENGINEER Wildfire at the same time, they were a bit apprehensive.

These fears were erased after Engineering Manager Steve Maryniak discovered that Pro/ENGINEER Wildfire's user-friendly graphical interface delivered a virtually seamless transition. He explains, "We found Pro/ENGINEER Wildfire to be very easy for modeling 3D assemblies."

Newfound Benefits of 3D Design

"We were a pretty traditional design shop," states Maryniak. "We made a lot of physical prototypes to test our design concepts. Our approach is now evolving to take a lot of that concept work back to the screen."

A Pro/ENGINEER 3D model makes it easy to visualize the design and improves communication, especially with non-technical departments. With a realistic model, the marketing department is able to get input early and ensure that the design is truly capturing the market need. In addition, design reviews are more efficient because the 3D data makes it a lot easier to clearly communicate the design intent.

Aesthetics

How stylish is your lawnmower? You may think that function always beats form in the lawnmower industry, but Maryniak reports that styling is becoming more and more important with consumers.

"Our customers expect the product to work, and for it to be durable. But styling makes an initial impact, and it is a differentiator for Hayter; the product that catches the buyer's attention has the advantage. With Pro/ENGINEER Wildfire, we have more flexibility in the early stages of design. This lets us be more responsive to initial feedback from the sales force, dealers, and distributors."

Pro/ENGINEER Interactive Surface Design Extension (ISDX) allows Hayter to consider the aesthetics of a product and develop an eye-catching design. Pro/ENGINEER ISDX makes it easy to create the right look to catch the consumer's attention. And it is 100% integrated with the other Pro/ENGINEER 3D design functions, so it simplifies the design of a stylish mower.

Smoother Manufacturing and Assembly Process

With Pro/ENGINEER, design changes are now automatically reflected in the manufacturing model without having to make manual updates, saving time and reducing errors. The engineering team also found that better data accelerates the toolmaking process, resulting in a 25-30% improvement in tooling lead times.

Hayter uses PTC's assembly management tools to reduce tedious tasks like building the bill of material (BOM). Plus, assembling the parts in 3D enables Hayter to check for interferences before problems show up in a physical prototype. In one instance, Maryniak recalls how engineers hadn't accounted for the height of the head of a bolt attaching the cover to the mower's belt. "Interference checking prevented us from making some fairly basic, yet costly errors."

Hayter is also using Pro/ENGINEER to create better support materials. The parts book now includes exploded images taken directly from Pro/ENGINEER; previously, drawings were produced separately using Adobe Illustrator and managed independently of the design data.

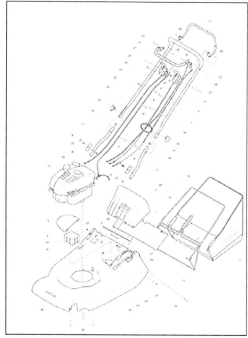

Lawnmower exploded view for operators manual

The Future

Using PTC solutions has enabled Hayter to reduce front-end design time by more than 300 working hours, and cut the time to manufacture the mowers' aluminum decks by two months. By sharing its 3D models with sales and marketing, the design team has also been able to improve collaboration throughout the product lifecycle. The company's commitment to product excellence has won the loyalty of its customers; over 80% of customers consider replacing their aging machines with new Hayter models.

Hayter's superior products and loyal customer base recently caught the eye of The Toro Company, the $1.6 billion (US) industry giant, which was looking to expand its UK presence. In February of 2005, Toro acquired Hayter, recognizing the company's well-established base of engineering and its commitment to product excellence.

And a bonus for the Hayter engineers: Toro has been a Pro/ENGINEER user for years, and the design teams from both companies have started working together to leverage each other's strengths.

Appendix

Checklist for 2D to 3D Conversion			
Name:			
Company:			
Product:			
Requirements	Status	Comments	To Do
Hardware			
Existing			
New			
Software			
Cost			
Personnel			
Existing			
New hires			
Training			
In-house			
External			
Production			
Internal			
External			
Timeline			

| Name: |
| Company: |
| Product: |

2D CAD design Survey for
Moving from 2D to 3D CAD for Engineering Design
Challenges and Opportunities

Question	*Answers*
What product does your company produce?	
How long has your company used 2D CAD?	
What 2D CAD system do you use in the design process?	
Do you manufacture in-house?	
What CAM system does your company use?	
What CAD systems have you demoed?	
How many people in your company use CAD in their daily assignments?	
What would be the motivating consideration in adopting a 3D CAD system?	
Would your workforce resist or embrace the change?	
How much does the cost of 3D software figure into your decisions on design software?	
How much does the cost of transition (training, etc) figure into your decisions on design software?	
What impression have you had when dealing with 3D CAD companies and their sales staff?	
What would be the primary impetus to you deciding to move to 3D design?	
What assistance would you like to see from a 3D CAD company?	
What other concerns do you have?	
Would you mind being contacted for a short phone conversation about this material?	Phone number:

Thank you for your participation. Please return this to lgl@cad-resources.com

Index

2D CAD iii, viii, 1-3, 6-7, 19, 21-24, 26
2D design data 14, 91
2D drawings 6, 22, 36, 90-91, 119, 122
3D assemblies 4, 159, 205
3D CAD iii, viii, 1-7, 9-25, 27-29, 31-37
3D design data 18
3D geometry 91
3D modeling 2, 18, 21
Analysis 2, 10, 14, 23, 32, 40, 58,
    144, 155, 157, 173, 178, 183, 188-189
Animation 2, 7
Annotations 48
ASME 20, 40
Assemblies 2, 4, 22-23, 52, 57-58, 60, 64,
    92, 159-160, 168, 170, 181, 192, 205
Assembly constraints 75
Assembly mode 52, 62
Assembly model 57, 71, 73, 85, 181
Associativity 23, 28, 48, 63, 176, 188, 201
AutobuildZ 92-93, 119-140
Auxiliary views 125
Balloons 86-87
Benchmark 12, 13, 16, 157, 205
Bi-directional 6, 18, 22, 63
Bill of materials 59, 87, 182, 192, 206
BOM 2, 58-60, 73, 85, 87-88, 180, 192, 206
Bottom-up design 57, 61-62
Cabling 157, 184
CAD iii, 1-7
CADTRAIN 20, 145, 152
CAD/CAM iii, 162, 176
CAD vendors 158, 161
CAE viii, 4, 188
CAM iii, 2, 4, 18, 22, 54, 56, 63, 151,
    157, 162, 175-176, 202, 209
CBT 20, 145, 147, 152, 155
Centerline 41, 47, 113, 123
Chamfer 32-33, 37, 95, 108, 114
Child 28-29, 32
Community college 142, 149-150
Computers 13-14, 149, 163
Concurrent engineering processes 4
ConnectPress 155
Construction plane 101
Consultants 141, 144
CNC 157, 181
Cut 24, 28, 62, 109-110, 132-135, 181
Data translation 91, 176, 188-189
Datum plane 95, 97, 99, 101, 103, 112, 129

Design 1-4, 6-7, 9-25, 27-28, 32-34, 61-62,
    64-66, 116, 141, 146, 162, 173, 175,
    181, 188, 192, 196
Design intent 21, 32, 34, 40, 147, 160,
    183, 203, 206
Design layout 64, 185
Design size 27
Designers 2-3, 5, 11, 13, 18-19, 36, 62,
    91-92, 142, 146, 160, 164, 175,
    177, 179, 184-185, 188, 201, 203
Detail views 45
Detailing 19-21, 36, 51, 61, 66, 181
Digital models 184
Dimensioning 23-24, 35
Dimensions 24-25, 37-38, 40-44, 47-48,
    50, 61, 115, 123, 160, 182, 203
Documentation package 6, 22-23
Drafters 18-19, 146
Drawings 6-8, 17, 22-23, 34, 36, 40, 48,
    58-61, 90-91, 116, 122-124, 160,
    180-183, 192, 207
Drawing mode 48, 52, 128
Drawing scale 125-126
ECO 10, 18-19, 23, 28, 34, 36, 56-57,
    93, 162
Editing 28, 41, 178
Engineers 2-4, 7, 22, 32, 62, 91,
    141-142, 146-148, 163, 169-170,
    183-184, 192-196, 200, 202, 205-207
Exploded view 71, 207
Extrude(d) 25, 67, 109, 112, 128-129,
    132, 134
Extrusion 24, 28, 32, 65, 102, 128-130
FEA 32
Frotime 153
Graphics card 165-169, 171
Hardware 4, 13, 16, 141, 144, 149, 157,
    163-171
Hole(s) 34-35, 41, 44, 46-47, 95, 110-111,
    128, 135-138, 180-181
IGES 91, 119, 159
Implementation iii, 9-10, 12, 15-17, 141, 169
Imported drawing(s) 96, 123-125
Imported geometry 97, 99-100, 106, 111,
    113, 178
IT 163, 170, 190-191, 193
Large assemblies 2, 14, 166
Legacy data (designs) 1-3, 10, 14, 18, 20, 91,
    93, 96-97, 104, 139, 157
Linux 165, 167, 172
Machined parts 14

Machining 2, 36, 54-56, 162, 175-176, 181
Manufacturing viii, 2, 4, 7, 14, 18-23,
  40-41, 44, 54-58, 63, 116, 141-142,
  157, 161-162, 173-176, 178, 181,
  190-192, 194, 200-202, 206
Manufacturing department 141
Manufacturing process 19, 63, 202
Material(s) 14, 23-24, 60, 77, 109, 182
Mechanism 2, 18, 32, 58, 117, 141, 157
Memory 13, 167-170
Mirror 34-35, 196, 198
Mold design 9, 14, 22, 173, 180-182
Monitors 14, 163
Multi-axis machining 2, 175
NC 4, 18, 22-23, 32 36, 63, 157, 162,
  175-176, 180-181
Networking 170-171
Notes 40-41, 44, 46-49, 61, 123, 167-168,
  180, 182
Online training 3, 142, 145
Orthographic 22-23, 30, 36, 119,
  125-126, 128
Parameter(s) 12, 32, 52, 59-60, 62, 64,
  73, 77, 83, 86-87, 157
Parametric (modeling) 6, 21-22, 28, 32,
  40, 47, 48-49, 56, 62, 93, 96-97, 119-120,
  128, 176-179, 181, 202-203
Parent 25, 28, 32, 36, 40, 45, 95
Parent-child 32
Pattern(s) 34, 137, 179, 182
PDM 10, 14, 20, 159, 191-193
Piping 157, 185
PLM 14, 190-194
Processor(s) 165-169, 171
Product design 2, 19, 142, 157, 177-178,
  185, 188, 192, 195-196, 205
Productivity iii, viii, 2, 59, 141, 157, 164,
  170, 176, 185, 188, 205
RAM 164-171
Rapid prototyping 40, 173
Reference dimension(s) 23, 125-126
Relation(s) 32, 64, 79
Relationship(s) 28, 32, 62, 64, 178, 194,
  202-203
Rendering 2, 5, 23, 173, 186-187
Reorder 28
Reverse engineering 62, 183
Revision(s) 48, 57
Round 32, 179, 182
Scalability 1-2, 4
Scale 23, 49, 125-126, 180

SDC 153
Section geometry 24, 65, 67, 78-79, 133
Section views 43
Selection 9, 12, 15, 33, 121, 132, 157-158,
  163-164, 169, 182, 185
Servers 170, 192
SFV ii
Sheetmetal 157
Simulation 55-56, 167, 173, 176, 178,
  180-181, 188-189
Sketch 24, 64, 92-93, 97-99, 101-104, 106,
  109-110, 112, 119, 130, 132
Sketcher 24, 97
Sketching 6, 25, 27, 92, 102, 129-130,
  133, 135
Sketching plane 27, 129-130, 133, 135
Software 1-4, 10, 12-14, 16, 28, 32, 119,
  141-145, 147, 149, 151-154, 157-158,
  160-165, 169-170, 185, 192, 195-198,
  200, 205, 209
Software selection 157-158
Standard component (part) 61-62
Student edition software 152
Surface(s) 9, 24, 28, 32, 104, 129-130, 133,
  135, 157, 173, 176-180, 183, 195, 197-198,
  202-203, 205-206
Surface design 157, 177, 183, 195, 197,
  205-206
System requirements 165-168
Tables 64
Technical illustration 7, 57
Title block 48-49, 61-62, 77
Tool design 179-180
Tooling 23, 58, 179, 206
Top-down design 61-62, 66
Training 1-4, 11-14, 16, 18, 20, 91, 141-149,
  152-153, 155, 157-158, 187, 198
Transition 1-2, 7, 12, 15-16, 18-20, 92,
  141-142, 144, 164, 171, 205, 209
Tryout edition 152
UNIX 165, 167, 172
Usability 2
User groups 154
VARS 144, 158, 171
View 7, 22-26, 28, 34, 36, 40-41, 57, 61-62,
  66, 87, 89-90, 92, 95-97, 115, 119, 124-129
Vista 169, 171
WEB 16, 120, 143, 146-148, 151, 153-155
Windows 3, 142, 152, 165-168, 171-172
Workflow 19, 21, 57, 102, 121, 188, 194

www.ingramcontent.com/pod-product-compliance
Lightning Source LLC
Chambersburg PA
CBHW080407060326

40689CB00019B/4166